Naval Engineering

NAVAL ENGINEERING

The Principles of Fire Protection

JACK HOPSON, JR.

3 Pillars Publishing, LLC
1773 Chase Pointe Circle
Unit 518
Virginia Beach, VA 23454
Phone: 1 (757) 333-3466

Published by 3 Pillars Publishing 06/2019

ISBN: 978-0-578-44162-7

Library of Congress Control Number: 2019900019

Printed in the United States

Table of Contents

Acknowledgement		vii
Preface		ix
Introduction		xi
Chapter 1	Fire Protection Engineering	1
Chapter 2	Shipboard Firefighting Organization	27
Chapter 3	Fundamentals of Shipboard Fire Protection	61
Chapter 4	Fire Protection Systems	101
Chapter 5	Submarine Fire Protection Engineering	149
Chapter 6	Shipboard Firefighting Tactics	167
Chapter 7	Aviation Fire Protection Engineering	205
Chapter 8	The Fundamentals of Aviation Fire Protection	225
Chapter 9	Aircraft Firefighting Tactics	243
References		263
Appendix I	Glossary	265
Appendix II	Abbreviations and Acronyms	283
Appendix III	Index	287

Acknowledgement

To the extent this book may contain text in the public domain and the Author makes no claim of ownership. The Author is credited with text compilation and editing. Photographs were taken by the Antonio P. Caliz and Miguel A. Vazquez and release to the Author.

Preface

Naval fire protection engineering took flight when one of the most serious fire disasters in naval history occurred on July 29, 1967 aboard the *USS Forrestal* (CVA 59) while operating off the coast of Vietnam. The fire was started by the accidental launch of a missile from one of its aircraft as they prepared to launch off the flight deck. This fire was a total catastrophe as the fire and fuel spread across the flight deck causing explosions in bombs and fuel tanks of aircrafts. These explosions caused holes in the flight deck which produced fires from burning fuel spilling to the decks below the flight deck. The crew fought the fire for over 24 hours to extinguish the fires below the flight deck.

The losses from the *USS Forrestal* fire incident included 134 deaths, 161 injured, over 20 aircrafts destroyed and repairs costing approximately $72 million dollars (equivalent to $528 million as of 2018). The repairs to the ship took over 2 years to complete. A detailed investigation report determined that most of the ship's experience firefighters were killed on the flight deck during the explosions and many of the surviving firefighters were unfamiliar with firefighting procedures and how to operate the firefighting equipment. The lessons learned from the horrific incident lead to creation of flight deck aqueous film-forming foam flight deck and deck edge sprinkling systems, modification in firefighting procedures, and improvements in fire hoses, fire nozzles, firefighting clothing, and personal protective equipment. Also, multiple exits were added to new built ships and egress markings were created from living quarters and below deck compartments. The Navy spent a great emphasis on firefighting and damage control readiness through providing specific training and updating firefight systems and equipment.

The Naval Research Laboratory (NRL) has maintained strong participation in interdisciplinary research and development program for new theories, systems, materials, and guidelines in fire protection and damage control. Many of these tests are conducted onboard the Navy's fire test ship known as the *ex-USS Shadwell*. The Navy has special concerns for fires in semiconfined compartments that are below decks in surface ships and in totally enclosed compartments found on submarines where fires function differently from most structural fire situations in the civilian sector. Such Navy compartments contain state-of-the-art equipment that must continue to operate under adverse condition and these types of areas must rely on crew firefighting abilities. NRL's developments continuously modernized Navy's fire protection to provide effective firefighting methods utilizing minimal manpower. The NRL advances have made great contributions to the Navy as well as impacting the civilian sector.

Introduction

A fire is like a dancing and roaring pride of lions that devours every living thing in sight without any remorse. I came face to face with these horrific creatures on my 31st birthday onboard the *USS Rushmore*. The *USS Rushmore* pulled into Pearl Harbor Naval Base for the weekend before heading on deployment. Repair Division leaders had gather in the Machine Shop to discussed the day's accomplishments and the weekend worklist. Near the end of our meeting, one of my Sailors, Eli'jah Parker ran into the shop, grabbed a portable fire extinguisher and ran out the shop leaving the leadership in an astonished manner. I quickly snapped out of it and hurled forward through the shop door into the passageway and followed Parker. He opened the door to the weather deck and turned towards the smoke deck. As I followed him, I thought briefly, "It must be another fire in the cigarette butt can." When I got to the smoke deck, a sault of lions was roaring at Parker and bystanders. Flames and heavy smoke was dancing out of an intake vent. As the ship's fire marshal, my instincts immediately took over. I ordered all personnel to evacuate the smoke deck and directed someone to report the fire to the Quarterdeck. One of the bystanders was an electrician, I directed him to secure ventilation and power to the intake room.

As soon as the bells started ringing over the ship's announcement system, I had just finished breaking out a firefighting hose and nozzle. Parker continued to spray extinguishing agent at the lions but they started roaring louder and louder as the fire extinguisher's agent diminished. The lions overtook Parker once his extinguisher was emptied. He began coughing from smoke inhalation that was thrusting towards him. After crouching down and heading towards Parker. I immediately grabbed him by his collar and pulled him out of the smoke. I asked him, "Are you ok?" He replied with a nod that he was well. I directed him to pick up the nozzle as I carefully charged the fire plug. Once the hose was fully charged, I got on the fire hose to give Parker support and we began firefighting efforts.

We continued fighting the fire, until we were relieved by the In-port Emergency Team. Once we were relieved, I immediately noticed that the ship's leadership was standing around observing the event. I knew everyone was concerned but they were in harm's way, so I ordered every bystander to evacuate the scene. I roved around to ensure that the ventilation system was safe and all boundaries were set. I briefly thought, "There is no way that we are going to let this fire spread. Not on my watch." The In-port Emergency Team fought the lions fiercely and their efforts finally overcame the lions. The fire was extinguished and the team discovered bails of rags in intake room. It took several hours to completely overhaul the intake room and the bails of rags. The investigation concluded that cigarette butts were sucked into the intake room causing the fire. Although this could have been catastrophic, this event was one of the most exciting birthdays I can remember. I was very proud that our fire protection equipment worked and our continuous training paid off by us working together as a team to effectively extinguish a fire. My experience as a fire protection engineer has fueled my passion to write this book, I strongly believe that preventing, controlling, and mitigating the effects of fires is in keeping with the history of protecting our naval assets.

Fire at sea has always posed an absolute danger to U.S. Navy ships. Shipboard firefighting is more than putting the "wet stuff on the red stuff," although it is true that some tactics used in structural firefighting can be applied in many cases of shipboard firefighting; however, there are tactics specific to shipboard scenarios that must be learned. In U.S. Navy ships, the fire hazard is magnified by the threat of explosive weapon warheads and propellants, but the principal threats on Navy ships are fires produced in the machinery and auxiliary spaces. Throughout history the U.S. Navy has dealt with fire protection challenges by demanding the most from existing firefighting systems through organization and training as well as implementing new technologies. The application of specialized naval firefighting systems such as heptafluoropropane and water mist systems are examples of the constant improvements pursued by the U.S. Navy in the safety and survivability of its vessels, aircraft, and crews.

Firefighting is an "All Hands" effort and it greatly depends on all personnel's knowledge of firefighting tactics and their ability to operate fire protection systems and equipment to ensure naval ship's survivability. *Naval Engineering: The Principles of Fire Protection* provides a general understanding of fire protection engineering aboard naval ships. The fundamentals of fire protection are compiled for the Military Occupation Specialty rating known as Damage Controlman and Aviation Boatswain's Mates as well as for fire safety and fire protection engineering professionals. The reader will get basic principles of organizational roles, functions of fire protection systems and equipment, firefighting training, tactics and readiness.

CHAPTER 1

Fire Protection Engineering

Fire protection engineering is applying science to engineering principles to protect people, property, and their environments from the harmful and destructive effects of *fire* and smoke through fire detection, suppression and mitigation. Fire safety engineering is another element of fire protection engineering which focuses on human behavior and maintaining a dependable environment for evacuation from a fire. This discipline includes the following:

1. Fire detection – fire alarm systems and brigade call systems.
2. Active fire protection – fire suppression systems.
3. Passive fire protection – fire and smoke barriers, space separation.
4. Smoke control and management.
5. Escape facilities – emergency exits, fire lifts, etc.
6. Building design, layout, and space planning.
7. Fire prevention programs.
8. Fire dynamics and fire modeling.

9 Human behavior during fire events.
10 Risk analysis, including economic factors.
11 Wildfire management.

Fire protection engineers assist *architects*, designers and developers with evaluating life safety and property protection goals for buildings and facilities by identify risks and design safeguards that aid in preventing, controlling, and mitigating the effects of fires.

The origin of fire protection engineering dates back to 64 B.C. during the reconstruction of *Rome* after a disastrous fire that almost left the city in ruins. Roman Emperor *Nero* commanded that the city be rebuilt utilizing fire protection methods such as the separating buildings materials and building with non-combustible materials. As fire protection engineering evolved over the centuries, the field was included in civil, mechanical, and chemical engineering disciplines during the Industrial Revolution era. Fire protection engineers during this era concerned themselves with developing methods to protect large factories, *spinning mills* and other manufacturing properties. During the latter part of the 19th century catastrophic conflagrations such as the great fire of Chicago and other major cities fires was the motivation to organize the discipline, define practices and conduct research to support innovations for fire protection. In 1903, the first academic degree program began at Armour Institute of Technology which later merged with Lewis Institute to form Illinois Institute of Technology. In the 20th century, fire protection engineering emerged as a unique profession separating from the *civil, mechanical* and *chemical engineering* disciplines. This emergence led to the development of the "body of knowledge," specific to the creation of the Institution of Fire Engineers (IFE) in the United Kingdom and the *Society of Fire Protection Engineers* (SFPE) in the United States.

Fire protection engineers must complete a formal *education* and upon graduation they must *continue their professional development* and maintain their competency through experience. This formal education usually includes studies in mathematics, physics, chemistry, and technical writing. Fire protection engineering is a professional engineering discipline that focus its students on acquiring proficiency in *material science*, *statics*, *dynamics*, *thermodynamics*, *fluid dynamics*, *heat transfer*, *engineering economics*, *ethics*, *systems in engineering*, *reliability*, and *environmental psychology*. Other classes in the fire protection engineering may include studies in *combustion*, *probabilistic risk assessment* or *risk management*, the design of *fire suppression systems*, the application and interpretation of model *building codes*, and the measurement and simulation of fire phenomena. *Society of Fire Protection Engineers* (SFPE) recognizes the following schools that offer Bachelor or Master of Science degrees in fire protection engineering:

1. Carleton University (Canada).
2. California Polytechnic State University (US).
3. Case Western Reserve University (US).
4. Ghent University (Belgium).
5. Hong Kong Polytechnic University (Hong Kong).
6. International Master of Science in Fire Safety Engineering (Europe).
7. Karlsruhe Institute of Technology (Germany).
8. Lawrence Technological University (US).
9. Luleå University of Technology (Sweden).
10. Lund University (Sweden).
11. State Key Laboratory of Fire Science (China).
12. Stord Haugesund University College (Norway).
13. Technical University of Denmark (Denmark).
14. Ulster University (Northern Ireland).
15. Universidad de Cantabria (Spain)
16. Universidad Pontificia de Comillas (Spain).
17. University of Canterbury (New Zealand).
18. University of Coimbra (Portugal).
19. University of Edinburgh (Scotland).
20. University of Greenwich (England).
21. University of Leeds (England).

22 University of Maryland, College Park (US).
23 University of New Haven (US).
24 University of Queensland (Australia).
25 University of St. Thomas (US).
26 University of Waterloo (Canada).
27 University of Western Sydney (Australia).
28 Worcester Polytechnic Institute (US).

Naval Fire Protection Engineer

All seagoing vessels are unique as nearly every compartment of the ship may require a specialized type of fire protection system. What makes U.S. Naval vessels unique is that they function like miniature floating cities. Naval vessels functions as fuel storage farms, high voltage power plants, airports, hotels, restaurants, and warehouses with all kinds of commodities. These vessels possess hazards in the engine rooms, generator rooms, control rooms, ammunition magazines, flight decks, hangar bays, and kitchen areas potentially require different fire protection methods. Majority of firefighters not experienced in shipboard firefighting tactics may assume that fires on ships can be extinguished in the same manner as a structure fire. Shipboard fires must be handled differently as there is an obstacle course throughout the vessel. Another issue is ships are made of steel which makes a compartment fire is similar to an oven. Heat is radiated back into the center of the compartment, unlike a structure fire, which tends to absorb the heat. With all of these hazards identified, the fire protection engineers must be knowledgeable of the fire protection methods and must be able to train shipboard firefighting teams to effectively extinguish shipboard fires.

The fire protection engineers in the United States Navy are known as a damage controlman and aviation boatswain's mate; however, the aviation boatswain's mate role as a fire protection engineer will be discussed in chapter 4. The damage controlman rating is a military occupational specialty (MOS) in the engineering field that covers a wide variety in the subject matter of fire protection and emergency management. A fire protection engineer will perform preventive and corrective maintenance to fire protection systems, fire protection equipment and emergency repair equipment on the organizational and intermediate maintenance levels. They shall plan, supervise and perform tasks necessary for firefighting, battle damage repairs, preservation of ship's stability, and defensive measures against chemical, biological, radiological, and nuclear (CBRN) warfare. They train, instruct and coordinate firefighting and emergency response teams in techniques of firefighting, battle damage repair and CBRN warfare defense.

As Fire protection engineers advances through the military ranks, they will find great responsibilities in military and technical leadership. Their responsibilities in military leadership are equivalent to other military occupations in the enlisted ranks (E-4 to E-9) as they all become military leaders and technical specialists. The fire protection engineer responsibilities for technical leadership are very significant and unique as it correlates their expertise to their ability to lead personnel during emergency situations. They hold key positions in the emergency response organization during firefighting and battle damage repair situations. These types of emergencies are "all-hands" evolutions with fire protection engineers leading the way.

In order to coordinate the efforts of personnel for the successful control of fires and battle damage, the fire protection engineer must possess great leadership as well as knowledge, skills and abilities in the discipline of fire protection engineering. Organization and teamwork are the keys to successful fire extinguishment and battle damage repairs. Strong leadership keeps the organization functioning by utilizing effective teamwork to meet the following goals:

1. Preserve or restore the ship's watertight and fume tight integrity.
2. Preserve or restore the ship's stability, mobility, and offensive ability.
3. Conduct emergency repairs to ship's systems and equipment.
4. Extinguish and limit the spread of fires.

5 Remove and contain the spread of contamination from chemical agents, biological agents, toxic gases, and nuclear radiation.
6 Care for wounded personnel.

In order to achieve the above organizational goals, the emergency response organization must work toward these three basic objectives:

1 Take all practicable preliminary measures to prevent fires and emergencies.
2 Extinguish fires and localize damage as it occurs.
3 Accomplish emergency repairs as quickly as possible, restore shipboard systems and equipment to operation, and care for injured personnel.

The ship's ability to perform its mission will depend upon the effectiveness of its emergency response organization. This organization has the same objectives whether the country is at peace or at war. To meet these objectives, the emergency response organization needs to accomplish the following actions:

1 Preserve the ship's stability and buoyancy by maintaining fumetight and watertight integrity.
2 Maintain the operational capabilities of vital shipboard systems.
3 Remove the effects of fire and explosion through prevention, isolation and extinguishment.
4 Remove the effects of chemical, biological, radiological and nuclear contamination through detection and isolation.
5 Prevent personnel casualties and administer first aid to the injured.
6 Conduct rapid repairs to ship's structure, systems, and equipment.

The following are the firefighting and emergency response areas of responsibility:

1 The functional combination of all equipment, material, devices, and techniques that are used to prevent damage, minimize damage, and restore damaged systems, equipment, and structures that can occur during wartime or peacetime.

2 The passive defense against conventional, nuclear, biological, and chemical warfare.
3 All active defensive measures short of those designed to prevent successful delivery of an enemy attack by military means or sabotage.

Professional Development

A fire protection engineer will work in emergency preparation, fire safety, and fire prevention on a daily basis. During the routine preparation of preventing fires and emergencies, they will inspect and maintain emergency response equipment, firefighting systems and firefighting equipment. Fire protection engineers are assigned to firefighting and emergency response teams. They are required to familiarize themselves with the ship's systems, ship's layout and all aspects of fire protection and emergency response equipment, systems, and procedures. Eventually, they will qualify as on-scene leader (OSL) for the firefighting and emergency response teams. As they gain experience, they will help train the ship's personnel in fire protection systems, firefighting equipment, and emergency response procedures. Although firefighting and emergency response is an "ALL-HANDS" responsibility, fire protection engineers ensure that firefighting and emergency response readiness is kept at the highest degree.

Navy Schools

The professional development of fire protection engineers is very critical to U.S. Navy ships' survivability. They first receive their technical training through instructor-led and computer base training courses conducted at Damage Control School held at Training Support Center Great Lakes, Illinois shown in figure 1-1. This course is approximately 10 weeks long and includes the following curriculum:

1 Basic first aid.
2 Shipboard safety.

3 Blueprint reading.
4 Industrial training.
5 Record keeping.
6 Mechanical systems maintenance and repair
7 Introduction to computers.
8 Introduction to marine engineering.
9 Firefighting strategies.
10 Fire protection systems.
11 Portable fire extinguishing equipment.
12 Fixed extinguishing systems.
13 Fire science.
14 Applied firefighting methods.

As they pass through the military ranks, the journeyman-level to master-level fire protection engineers will go to several advanced technical training courses. The journeyman-level fire protection engineers will attend courses to learn gas free engineering. A program that is vital for certifying and authorizing personnel to enter into shipboard confined spaces and oxygen deficient rooms. Another important course for the journeyman-level fire protection engineer is the Foam Generation Course which teaches them operation and proper maintenance on the Aqueous Film-Forming Foam (AFFF) fire protection system. They also attend the CBRN Defense course taught at Fort Leonardwood Army Base in Missouri where they learn how to prepare, defend, and recovery from chemical, biological, radiological, and nuclear attacks and hazards. Once they complete this advanced course they will have earned the Navy Enlisted Code 756B where they become a shipboard chemical, biological, radiological and nuclear-defense (CBRN-D) operations and training specialist.

Figure 1-1. Damage Controlman School at Training Support Center Great Lakes

A master-level fire protection engineers is required to become the subject matter expert in the fire protection engineering occupation. They utilize experience gained through on-the-job training and training learned from attending Damage Control Assistant & Senior Enlisted Damage Controlman course taught at Naval Station Norfolk, Virginia or Naval Station San Diego, California. This advanced course teaches the managerial functions in support of firefighting, battle damage repair, emergency response, gas free engineering, CBRN warfare defense, and ship's stability and buoyancy programs. They will learn how to manage the preservation and maintenance of firefighting systems, firefighting equipment, emergency response equipment, and CBRN warfare defense systems and equipment. They will also learn how to support the damage control assistant (DCA), whose role will be discussed later, with organizing and training the ship's firefighting and emergency response teams through planning and executing firefighting and damage control exercises. Once this course is successfully completed, they will have earned a NEC U46A where they become a senior enlisted damage control program management and training specialist.

On-the-Job Training

A vital component in the professionally development of fire protection engineers is on-the-job training (OJT). This training method allows them to learn while performing their daily routines and tasks. Fellow colleagues and supervisors teach each other by training and sharing knowledge they have gained throughout the years. Master and journeymen-level fire protection engineers train the apprentice-level members how to perform preventive and corrective maintenance on firefighting systems and equipment. Another important OJT element is conducting drills and exercises which will help prepare journeymen and apprentice-level members to fill key positions within the firefighting and emergency response team organization. Conducting drills and exercises allows them to lead teams and get the team members working together as an effective unit.

Importantly, fire protection engineers can become better technicians by studying applicable technical publications. They will gain knowledge by enrolling in the following Non-Resident Training Courses:

1. *Damage Controlman (14057).*
2. *Blueprint Reading and Sketching (14040A).*
3. *Tools and Their Uses (14256A).*

Also they can improve their knowledge by studying technical manuals and other technical sources. The technical publications that they should become familiar with are as follows:

1. Manufacturers' Technical Manuals.
2. Naval Ships Technical Manual (NSTM).
3. Damage Control Books.
4. Repair Party Manual.

The manufacturers' technical manuals provide detailed information on the operation, maintenance, and repair of specific systems and equipment. These manuals are vital sources for fire protection engineers to ensure that systems and equipment are fully operational during emergency situations. Since there is a significant amount of manuals, the manuals listed below are a few examples:

1 *Fire Extinguishing System – Aqueous Film Forming Foam (AFFF) Systems Manual.*
2 *Fire Extinguishing System – Halon 1301 Fire Extinguishing Systems Manual.*
3 *P-100 Portable Pump Unit Technical Manual (Manufacturer) Operation, Maintenance, and Repair with Parts List.*

Fire protection engineers can use NSTMs to help them with their professional development. A complete library of these technical manuals should be available in each ship's engineering log room. They may be available on a single CD-ROM and they may even be available on a shared computer network. These technical manuals cover different aspects of firefighting and emergency response. Here are a few examples of the NSTM's that are used:

1 NSTM Chapter 555 volume 1 – Surface Ship Firefighting.
2 NSTM Chapter 555 volume 2 – Submarine Firefighting.
3 NSTM Chapter 074 volume 3 – Gas Free Engineering.

The Damage Control Book provides critical information needed during emergency situations. These manuals contain descriptive information, tables, and diagrams which can be used to isolate and minimize fires and battle damage. These manuals are furnished to all naval ships over 220 feet long and to some select smaller ships. Each manual is pertinent to the individual ship. The information provided covers the following six subjects:

1 Damage control systems (firefighting systems).
2 Ship's compartmentation (ship's rooms).
3 Ship's piping systems.

4 Ship's electrical systems.
5 Ship's ventilation systems.
6 General information.

The Repair Party Manual provides detailed information on the standard methods and techniques used in firefighting and emergency response. These procedures cover emergency communications, emergency power, and counter flooding. Fire protection engineers use this manual to aid them during firefighting and emergency response. Commanding officers, with the assistance of their CHENGs and DCAs, are responsible for ensuring that the standard repair party manual for their ship is completed with updated information. This manual is standardized specifically to each ship and will include the following information:

1. List of each firefighting and emergency response team area of responsibility, to include machinery spaces, storage spaces, and magazines (ammunition storage lockers).
2. Protective measures in respect to imminent air attack, surface attack, underwater attack, fire, collision, and CBRN attack.
3. Methods of investigating damage; necessary precautions and means of reporting damage.
4. Use of equipment for the following purposes: firefighting, flooding control, repairing damage in action (shoring, pipe patching, etc.).
5. Controlling CBRN contamination (monitoring, reporting, and decontamination of material).
6. Personnel casualty control (first aid and decontamination).
7. Primary and alternate methods of providing emergency service to vital systems by means of casualty power (emergency power), emergency communications, and jumpers to restore firemain system (firefighting water) or magazine sprinkling system (ammunition storage sprinkling).
8. Damage Control Central (DCC) location, equipment layout, communications, and personnel.
9. A chain of command diagram.

10 A secondary DCC description.

11 Firefighting and emergency response positions, including duties, functions, and responsibilities; subunits (where applicable); and required publications, plates, plans, and diagrams.

Because naval fire protection engineering covers a wide variety, firefighting and emergency response is an "ALL-HANDS" effort. Everyone on the ship from the commanding officer down to the newest crewmember plays a critical role. Training is essential for effective firefighting and emergency response teams and it can be accomplished in several ways. Each personnel may learn through schools, technical manuals, OJT training, shipboard training lectures, drills, and exercises. There are a number of Navy schools that trains personnel in firefighting and emergency response. The DCA and Training Officer track the ships' training requirements and requests quotas to send personnel to schools. They can send personnel as individuals or as a complete firefighting and emergency response team to these schools. Members of the firefighting and emergency response teams train together in live-fire scenarios where they learn to work together as a unit. These schools cover basic and advanced firefighting and emergency response courses taught throughout the United States and overseas. These courses include Gas Free Engineering, General Shipboard Damage Control Wet Trainer, General Shipboard Firefighting, Advanced Shipboard Firefighting (Figure 1-2), and Aircraft Firefighting (Figure 1-3).

As mention earlier, fire protection engineers train the ship in firefighting and emergency response procedures. A master-level fire protection engineer should have the ability to develop firefighting and damage control lectures and scenarios used to train all shipboard personnel. Firefighting and damage control training provides a means to increase individual or team's proficiency in operating portable and installed firefighting equipment. Training will also increase knowledge of specific firefighting tactics and damage control procedures. Adequate training allows personnel to complete firefighting and damage control tasks in an expeditious routine.

Figure 1-2. Advanced Firefighting Training at Firefighting School Mayport, FL.

Figure 1-3. Aircraft Firefighting Training at CASI Fire School Rota, Spain

The key to a successful training program is to develop a self-sustaining training capability through the use of shipboard training teams such as the Damage Control Training Team (DCTT). Training resources such as Afloat Training Group (ATG) and Firefighting Schools are used to build this capability. These two resources train the training teams and provide them with tools to effectively train the ship's firefighting and emergency response teams. Training is the key to enhanced firefighting and damage control readiness.

Scenarios

Experience has proven that integrated training scenarios provide good measures for training teams to conduct efficient exercises and drills. The ultimate goal is for the ship's training teams to attain self-sufficiency and to maintain proficiency by conducting realistic, safe, and progressive scenarios designed to meet specific training objectives. To be effective, training must be scheduled and conducted beyond the basic training phase and continue throughout the entire operating cycle.

Bills, Directives and Reports

Senior fire protection engineers must become very familiar with several bills, directives, and reports. They also are responsible for assisting with maintaining and updating these documents. These documents are used to govern the crew's actions before or during emergency circumstances. It may be necessary for them to provide input to the DCA by recommending that the most qualified personnel are holding key positions for any emergency situations. The following documents are important to the ship's survivability:

1. Battle Bill - Firefighting and Emergency Response Organization.
2. Rescue and Assistance Bill - organization and procedures when assisting other ships.
3. Cold Weather Bill - Preparing the ship for entering cold weather operations.

4 Toxic Gas Bill - organization and procedures responding to toxic hazards.
5 Training Records - Records of qualifications, training sessions, and schools.
6 Equipment Test and Inspection Reports - Record testing, inspecting, and maintaining firefighting and emergency response equipment and systems in accordance to the planned maintenance system.

Key Roles and Responsibilities

All members of the ship's company should know their firefighting and emergency response (damage control) responsibilities and realize the importance of fighting a fire and responding to any causality that may affect the ship's stability and personnel safety. The importance of efficient emergency response cannot be overemphasized. Maximum readiness can only be achieved by a firm program motivated by effective and dynamic leadership. The success of this program will be determined by enthusiastic, well-trained, and determined officers, subject matter experts, and crewmembers from all departments on board the ship. While all areas will not be completely covered, the basic responsibilities of key individuals from the newest crewmember to the commanding officer will be covered.

Commanding Officer (CO)

The CO is overall responsible for the survivability and safety of the ship and crewmembers. They must maintain a state of maximum effectiveness for war or other required services. The CO shall immediately devote every effort to prepare their command for further obligation after a battle or action by repairing as much damage as possible. To ensure maximum readiness the CO must ensure that the command is adequately trained. Training must continuously be done through lectures, schools, and exercises in all aspects of firefighting and emergency response.

The CO should be fully aware of all of the ship's vulnerabilities such as all degraded or nonoperational firefighting and emergency response systems and equipment. The CO shall ensure that the fire protection engineers immediately correct all shortfalls and defects.

Executive Officer (XO)

The XO advises the commanding officer on the status of the ship's firefighting and emergency response in regards to ship's survivability readiness. The XO must be intimately familiar with all firefighting and emergency response evolutions, and this includes supervision of all actions related. The XO depends on key personnel to help them carry out the following requirements:

1 The XO ensures that the damage control organization is conducting firefighting and emergency response training for the crewmembers.
2 The XO leads the DCTT in conducting firefighting and emergency response training scenarios.
3 The XO must maintain ship's readiness to combat all casualties, hazards and damage that threatens the ship.
4 During fires or emergency situations on submarines underway, the XO assumes the duties as man in charge at the scene also known as the on-scene leader (OSL).

Engineer Officer

The engineer officer or chief engineer (CHENG) is also known as the damage control officer (DCO). The DCO supports the commanding officer by conducting the following duties and responsibilities:

1 The operation and maintenance of the ship's main propulsion plant, auxiliary machinery, and piping systems.
2 The control of battle damage.
3 The operation and maintenance of electric power generators and electrical distribution systems.

4 The repairs to the ship's structure.
5 The repairs to material and equipment of other departments that are beyond their capacity but within the capacity of the engineering department.
6 Maintains the ship's hull integrity, machinery, and electrical system in battle readiness.
7 Supervises firefighting efforts by ensuring that the ship's fire bill is adequate. Assign and instruct personnel in accordance with the provisions of the bill.
8 Maintains interior communication equipment.
9 Restores engineering and ship control casualties.
10 Coordinates all naval shipyard work including all correspondence or communications on alterations or repairs to the hull and installed equipment.
11 Maintains the Planned Maintenance System and other operating and maintenance records.
12 Act as technical assistant to the XO to carry out CBRN warfare defense procedures.
13 Provides ship facilities, equipment, and key personnel to repair the hull and machinery. Ensures repairs to material and equipment of other departments that are within the capacity of the engineering department.
14 Organizes Damage Control Repair Station (DCRS) 5 Propulsion Plant Firefighting and Emergency Response Team (in accordance with the ship's battle bill.
15 Supervises the training of DCRS 5.
16 Assigns appropriate engineering crewmembers to other Firefighting and Emergency Response Teams in accordance with the ship's battle bill.

Damage Control Assistant (DCA)

The DCA is directly in charge of the fire protection engineers (damage controlmen). The DCA is a principle assistant to the CHENG and solely responsible for fighting fires, repairing damage, CBRN warfare defense and maintaining ship's stability and buoyancy. The DCA is the overall coordinator of firefighting and emergency response matters within the shipboard organization to include the responsibility of managing the ship's firefighting and emergency response training program. During emergency situations that involves fires and damage that threatens the ship's stability and buoyancy, the DCA will address the problem and may use any technical advice and assistance received from other departments. Any fires and damage that occurs while the ship is at general quarters will be handled as a battle casualty. The DCA will direct corrective actions by utilizing the DCRS in the vicinity of the casualty except on aircraft carriers in which the air officer will direct firefighting and emergency response for fires and damage occurring to aircraft or associated equipment on the flight deck or in the hangar bays. The DCA will carry out the following firefighting and emergency response (damage control) duties and responsibilities:

1 Prepares command approved directives in connection with all emergency response functions requiring coordination of departments.

2 Submits firefighting and emergency response training schedules to the command's planning board.

3 Prepares a firefighting and emergency response training syllabus and provide applicable instructors for training all the crewmembers.

4 Furnishes all standard firefighting and emergency response equipment to DCRS and to other designated locations throughout the ship. Conduct periodic inventories and inspections of the firefighting and emergency response equipment.

5 Assigns all fire protection engineers to various firefighting and emergency response teams in accordance with the ship's battle bill and manning document.

6 Conducts periodic inspections throughout the ship, accompanied by department heads to ensure that the firefighting equipment and

ship's watertight integrity is maintained. Ensure that all departments are maintaining a high degree of firefighting and emergency response (damage control) readiness through fire prevention and fire safety.

7 Ensures that the master copy of the *Damage Control Book* is updated whenever alterations are made to the ship.

8 Ensures that damage control markings, routes, stations, and labels are posted throughout the ship.

9 Ensures that emergency escape routes to weather decks are clearly labeled.

10 Maintains DCC with items to evaluate damage to the ship's hull and equipment and to make decisions to counteract the effects of such damages. Coordinate the actions of firefighting and emergency response teams and keep the commanding officer informed of major developments.

11 Specify routes for transporting injured personnel to battle dressing stations (BDS).

12 Ensures that an effective emergency response organization is always present for execution of each of the emergency bills.

13 Immediately informs the CHENG of any conditions or practices that reduces the firefighting and emergency response (damage control) readiness of the ship.

14 Organizes DCRS 1, 2, 3, 4, and 7 in accordance with the ship's battle bill.

15 Personally directs the training of DCRSs 1, 2, 3, 4, and 7, and DCC personnel.

16 Coordinates with all department heads to ensure that damage control petty officers (DCPO) are designated and trained to accomplish their assigned duties which will be explained in a later in this chapter.

17 Functions as gas free engineer (GFE), manages the gas free engineering program, and supervises all the duties of the gas free assistants (GFEA) and gas free petty officers (GFEPO).

Fire Marshal

The fire marshal is an assistant to the CHENG and aids the DCA with training the crewmembers on fire prevention, firefighting, emergency response and battle damage repairs. The fire marshal is normally the most senior fire protection engineer on the ship except for aircraft carriers. In this case the fire marshal is a limited duty officer or warrant officer with an extensive background in firefighting and emergency response. The fire marshal must be thoroughly familiar with all the following:

1 All firefighting, emergency response and battle damage repair procedures.
2 Operate all firefighting and emergency response equipment and systems.
3 Know all technical documents that will enable them to be successful of the ship's survivability.

The fire marshal conducts daily inspections throughout the ship, paying particular attention to good housekeeping, fire equipment, fire hazards, and safety hazards. The fire marshal creates a report with the fire hazards and recommended corrective actions and submits it to the DCA with copies given to the XO and appropriate department heads. A follow-up inspection will be conducted to ensure that corrective actions are being taken. The fire marshal shall also carry out the following duties and responsibilities:

1 Trains the ship's At-sea Firefighting and Emergency Response Team, Rescue and Assistance Teams, and all the In-port Emergency Teams (IET).
2 Trains ship's fire-watch teams and makes their assignments prior to ship entering into overhauls, major repairs, and maintenance periods.
3 Trains junior fire marshals to lead their shifts on duty days (24-hour shifts) during in-port periods.
4 Take charge at the casualty scene until relieved by the OSL for firefighting and emergency response team.

5 Keeps the DCA informed of the exact status of the emergency situation.
6 While in-port, the fire marshal is responsible for the supervision of all the IET. In this situation, the fire marshal reports directly to the Command Duty Officer (CDO).

Officer of the Deck (OOD)

The OOD is the leading member of the underway watch team located in the Pilot House and serves as the primary assistant to the commanding officer. The OOD shall be intimately familiar with the ship, its material condition, and established procedures for emergencies. The OOD should know and understand the correct course of actions, or options, for various emergency situations. The OOD should be able to quickly analyze situations and take prompt, positive, and correct counteraction. The OOD's ability to react properly and promptly will be directly proportional to their knowledge of the ship's capability, firefighting and emergency response procedures, and equipment available.

Command Duty Officer (CDO)

The CDO is designated by the commanding officer and represents the CO when he or she is off the ship during inport periods. This officer is eligible for command at sea and is the deputy to the XO for a prescribed period of time. The CDO will carry out the following duties and responsibilities:

1 Carries out the ship's daily routine in-port.
2 Carries out the duties of the XO during the temporary absence of that officer.
3 Advises and directs the OOD in matters concerning the general duties and safety of the ship.
4 Stays cognizant of the ship's position, mooring lines, or ground tackle in use.

5 Remains aware of the status of the ship's engineering plant and all other matters that affect the safety and security of the ship.
6 Takes all necessary actions until relieved by a senior officer in the succession of command in times of danger or emergency.
7 Relieves the OOD when necessary for the safety of the ship and inform the commanding officer when such action is taken.
8 During fires or emergency situations on submarines in-port, the CDO may assume the duties as man in charge at the scene also known as the on-scene leader (OSL).

Damage Control Training Team (DCTT)

The purpose of the DCTT is to train the ship's damage control battle organization to implement current fleet tactics and procedures. This training team must first identify and develop a process by which shipboard personnel can effectively fight fires and control sustained damage as individuals and work together as an organized team. This is based on unique properties and systems of each class of ship and the training team members' experience, knowledge, and aptitude. DCTT shall be organized to train, evaluate, and critique the Firefighting and Emergency Response teams.

DCTT is composed of qualified senior members of the ship's crew specifically tasked to ensure the ship's company maintains the highest level of battle readiness. Its members should be the most qualified and motivated personnel on board. The effectiveness of DCTT is proportional to the level of command support for a strong qualification program. At a minimum, DCTT members shall be qualified to the position that they are assigned to train and evaluate. When selecting team members proven leadership abilities, experiences, and formal training courses in firefighting and damage control should be considered. Members of the DCTT should include the following:

1 DCTT Leader
2 DCTT Coordinator,
3 DCTT Team Members (Evaluators, Trainers, and Safety observers)

DCTT Leader

The XO serves as the chairman of the command's planning board and team leader of DCTT. The XO will coordinate the planning and execution of the ship's training effort. As the DCTT Leader, the XO is responsible for the management of the training team and requires the following:

1. Be a member of the command's planning board and DCTT.
2. Formulates a training package tailored to specific integrated or individual functional area team training objectives.
3. Identifies training constraints, disclosures and simulations, and annotate the training package accordingly.
4. Presents proposed training packages to the commanding officer for approval.
5. Conducts briefs for each training event for training team members and the firefighting and emergency response team being trained.
6. Ensures the training team conducts a thorough safety walk- through ensuring conditions are safe for training before each training event.
7. Supervises the manner of the training event.
8. Conducts the training event debriefs.
9. Establishes feedback mechanisms to address deficiencies identified during exercises.
10. Identifies training shortfalls and lessons learned from each exercises.

DCTT Coordinator

The ship's senior fire protection engineer holds the position of DCTT Coordinator. A senior hull maintenance technician (welder), machinery repairman (machinist), or chief warrant officer may hold this position if a fire protection engineer is not available. The DCTT Coordinator will find it necessary to coordinate, develop, and conduct intensive training and scenarios such as a Class B fire in an engine room or mass conflagration. Developing comprehensive shipboard damage control training programs often requires creating drills that coordinates with other training programs such as the Engineering Training Team and other training teams. Building coordinated training scenarios are vitally important as a fire casualty can be the result of an engineering casualty. These coordination efforts run from simple to complex, depending on all the training objectives. The DCTT Coordinator responsibilities to the DCTT Leader are as follows:

1. Organizing all team training periods, developing training event plans, and making all preparations in support of the event execution.
2. Serving as overall manager of the training event briefs, performance, and debriefs.
3. Training of team members in the proper conduct of their duties as drill initiators, exercise observers, and safety observers. These duties also include the operational risk management (ORM) process.
4. Compiling the results of the training event and submit the event evaluation sheets along with the critique sheets to the DCTT Leader for review.
5. Acting as coordinator for all recommendations and feedback concerning the training team.
6. Plan, schedule and execute training on identified training shortfalls and lesson learned trends.

DCTT Team Members (Trainers, Evaluators, and Safety Observers)

Trainers, evaluators, and safety observers directly observe individual and team performance of the training event and some may act as initiators. Their duties include the following:

1. Conduct on on-sight observations and evaluations.
2. Conduct safety walk through and pre-drill checks.
3. Provide training/prompting as necessary to meet the training objective during exercises conducted in the training mode.

4 Normally provide prompting only as required to prevent disruption of the event timeline or for safety reasons during exercises conducted in the evaluation mode.
5 Provide immediate feedback to individual watch standers upon completion of the training event.
6 Provide a post-exercise debrief on observations noted, lessons learned, and recommendations for corrective actions.

This chapter introduces the scope of fire protection engineering and how vital the roles and responsibilities of fire protection engineers play aboard U.S. Navy ships. The required training needed for their professional development is critical in maintaining the highest level of readiness in naval ships' survivability. Even though firefighting and emergency response is an ALL-HANDS responsibility, fire protection engineers are recognized experts that conduct training to the crew members, maintain the equipment and systems. Commanding officers are overall responsible for the operation, survivability, and safety of the crew during peace and wartime situations. Commanding officers rely on all the key individuals' responsibilities in preparing the ship for maximum readiness. The next chapter will discuss firefighting and emergency response organization's duties and responsibilities.

CHAPTER 2

Shipboard Firefighting Organization

Firefighting and emergency response is vital to all ships and submarines in the Navy. If a ship or submarine is damaged during battle, sabotage, negligence, storm, or normal deterioration, all fires must be extinguished and damage must be repaired immediately. Every ship and submarine must be organized to accomplish critical repairs to return the ship to normal operations. This organization is accomplished through assigned jobs, training, instructions, use of diagrams, and efficient communications. The firefighting and the emergency response organization establish standard procedures for handling various types of fire and damage. It sets up training for these procedures so that every person will know immediately what to do in each emergency situation.

Firefighting and emergency response has various vital objectives, both preventive and corrective. The prevention as well as the fighting of fires has proven essential to survival of a ship in peacetime and combat. Efforts must be continuously made to reduce the damage resulting from fires through elimination of hazards, keeping firefighting equipment properly maintained and operational, as well as effectively training emergency response parties. The following basic principles must be observed to reduce ship fire hazards:

1 Properly stow and protect all combustibles and understand how various chemicals react with others.
2 Make frequent and thorough inspections of all spaces, to include equipment, piping, and electrical cables/connections.
3 Educate all personnel in the reduction of fire hazards.
4 Enforce fire prevention policies and practices. To minimize fires, COs must eliminate nonessential combustibles, replace (whenever possible) combustible materials and equipment with less flammable items, and limit the amount of essential combustibles carried.
5 Maintain the established material conditions of readiness.
6 Train all personnel in all aspects of shipboard firefighting and emergency response.
7 Maintain firefighting and emergency response systems and equipment in the best condition possible to ensure survivability.

The corrective aspect of firefighting and emergency response requires the organization to be able to restore the offensive and defensive capabilities of the ship promptly. The firefighting and emergency response organization consists of two elements: the administrative organization and the battle organization.

Administrative Organization

The firefighting and emergency response's administrative organization is part of the engineering department organization. Each of the ship's department has major obligations to the administrative organization through preventive maintenance responsibilities for firefighting and emergency response equipment, systems, and fixtures within the departmental spaces. Each department head shall ensure that firefighting and emergency response maintenance tasks are completed and they will ensure that all discrepancies are documented and corrected.

Department Heads

Adequate firefighting and emergency response readiness can only be achieved by participation of all departments aboard ships. It is essential that each department head carry out the following duties and responsibilities:

1. Conducts periodic inspections of department spaces in accordance with current planned maintenance system procedures.
2. Ensures that firefighting and emergency response equipment and fittings are maintained in their proper locations and fully operational.
3. Provides personnel for firefighting and emergency response teams and other assignments as required by the ship's organization and battle bills.
4. Ensures that departmental material and equipment are secured to protect against damage by heavy seas.
5. Provides an immediate report to the DCA of any deficiency in damage control markings, firefighting equipment, or material, and initiate corrective action.
6. Coordinates with the DCA in ensuring departmental personnel are trained in firefighting and emergency response procedures.

Division Officers

The division officer is responsible for taking all preliminary measures before damage occurs, such as maintenance on firefighting and emergency equipment, removal of fire hazards, upkeep of ship's stability by maintaining watertight and airtight closures. Division officers ensure that all equipment, closures, and markings under their cognizance are kept in the best possible condition. This is done by periodic inspections, adherence to planned maintenance system checks by the division's DCPOs and training of personnel within the division.

Damage Control Petty Officer (DCPO)

A responsible member within each division is assigned as the DCPO for that division. The DCPO is responsible for firefighting and emergency response functions of the division and related matters. Division Officers are responsible to their Department Heads in providing the DCA with a qualified person assigned as DCPO. The DCPO should have received formal training and be qualified before assignment. Assignments as a DCPO are normally for a period of six months but may vary depending on the class of ship. They must follow the direction of the DCA and fire marshal during their period of assignment. The DCPO is responsible for performing and understanding the following duties and responsibilities:

1 Understands all procedures of firefighting, emergency response and CBRN warfare defense.
2 Assists with training divisional personnel in firefighting, emergency response and CBRN warfare defense procedures.
3 Ensures all damage control markings, routes, stations, and labels are posted throughout all divisional spaces.
4 Inspects portable firefighting extinguishers and equipment in accordance with planned maintenance system.
5 Ensures all divisional spaces' emergency battle lanterns, door opening wrenches, firefighting station wrenches, and other emergency response equipment are in place and fully operational.
6 Ensures all rooms, piping, cables, firefighting and emergency response equipment are properly labeled and identified by color codes and photoluminescence.
7 Ensures safety precautions and operating instructions are posted in required divisional spaces.
8 Assists the Division Officer in the inspection of divisional spaces for cleanliness and preservation, and assist in the preparation of required reports.
9 Conducts daily inspections of divisional spaces for the elimination of fire hazards.

10 Performs such other duties with reference to damage control and maintenance of divisional spaces as directed by supervisory personnel.

Gas Free Engineering Personnel

Gas free engineering personnel are equivalent to Marine Chemist. They are qualified personnel who have been designated as the GFE, GFEA, and GFEPO. The GFE is normally the DCA but this position can be held by a chief warrant officer or master-level fire protection engineer. The GFEA is position held by the master-level fire protection engineer or hull maintenance technician. The GFEPO position is held by journeymen level fire protection engineer, hull maintenance technician, machinery repairman or as necessary, other engineering rates.

The GFE oversees the gas free engineering program and personnel. The program is a necessary measures needed to eliminate all the risk of fire, explosion, exposure to toxic substances, and shipboard atmosphere that may cause suffocation or asphyxiation. Gas free engineering personnel conducts test on the atmosphere and depending on the test results, they may have to install portable ventilation to remove toxic gases and increase the level of oxygen to breathable levels between 19 to 22 percent. These test and measures must be completed prior to allowing personnel to enter into ship's compartments that have been closed off for extended periods, compartments that are poorly ventilated, after fires have been extinguished, hazardous spills, and leaks. Once the GFE receive the test results from the gas free engineering personnel, the GFE certifies the compartment as safe for personnel if the atmosphere conditions are within specified safe levels. Additionally, GFE personnel test the shipboard atmosphere prior to all welding operations.

Damage Control Supervisor (DCS)

The DCS is personnel that stand hourly shifts at DCC during normal ship operations and emergencies. The DCS will normally be a fire protection engineer, hull maintenance technician or machinery repairman. The DCS will carry out the following duties and responsibilities:

1 Supervises the maintenance of any material condition of readiness in effect on the ship. This includes the responsibility to check, repair, and keep the various hull systems in full operating condition.

2 Reports directly to the OOD on all matters affecting the watertight integrity, ship's stability, or other conditions that affect the safety of the ship.

3 Reports to the DCA for technical control and matters affecting the administration of the shift.

4 Maintains a written log. The log entries will show the hourly readings of the firefighting water (firemain) system pressure and the number of firefighting seawater pumps (fire pumps) in operation. Make entries such as the ship's leaving port, anchoring, and entering port. Also make entries of special evolutions such as general quarters, emergency drills, and the setting of material conditions, the discrepancies reported, and the corrective actions taken.

5 Supervises the maintenance of the ship's damage control closure log. List all fittings and watertight closures that are in violation of the prescribed material condition of readiness.

6 At the end of each watch, obtains the ship's fuel and water report on which fuel and water tanks were filled and emptied during the watch.

7 Reports hourly to the OOD on the status of the ship's watertight integrity.

8 When the ship is at-sea, have the Fire and Security Patrol (Sounding and Security Watch) take and report measurements of all tanks at least once during each 4-hour watch. While in-port, take measurements at least once each day. In addition, have the watch check the material condition of readiness in their respective areas. Report any corrective action taken in this respect.

9 Notifies the OOD, DCA, and weapons department duty officer when the fire alarm board indicates that the temperatures of any ammunition lockers are above 105 °F.

10 Ensures that the master key to the DCRS is issued only to authorized personnel.

Battle Organization

The battle organization is structured to allow the ship to continue its tactical mission while responding to a casualty in a tiered approach. A tiered response known as conditions of readiness allows the commanding officer the ability to utilize their resources efficiently. The tiered response consists of four layers of response:

1 Condition IV- During in-port normal working hours, the Flying Squad will be the initial responders for emergency situations. After working hours, the IET will be the initial responders for emergency situations. The Flying Squad and IET must also be capable of effectively controlling flooding and its possible effects, as well as, any other condition described in the General Emergency Bill.

2 Condition III- When the ship is at-sea, the initial response to fires and emergencies will be controlled by the Flying Squad. The Flying Squad is able to quickly respond to causalities and determine the extent of damage. The rapid response team proceeds directly to the scene of damage while the Flying Squad members provides from designated DCRS. If the emergency actions required for more assistance or a change in threat level, the ship will set Condition II Damage Control (DC).

3 Condition II DC- This condition allows significant increase in firefighting and emergency response without totally disrupting the ships routine operations. During this condition a designated DCRS and BDS shall be set up for operation. If threat levels increase the commanding officer can order another DCRS to be set or have the ship set Condition I General Quarters.

4 Condition I General Quarters (GQ) or Battle Stations- The ship has its maximum capability to withstand and recover from damage. Ships may set general quarters upon notification of a fire, but the

tiered response of using the Flying Squad and Condition II DC provides more flexibility. The commanding officer retains the option of setting GQ at any time.

The firefighting and emergency response's battle organization includes DCC, various DCRS, and BDS. The organization may slightly vary depending on the class of ship. The variance will depend upon the size, type, and mission of the ship; however, firefighting and emergency response team organizations have the same basic principles. These basic principles are as follows:

1 Ensures that all personnel within the organization are highly trained in all phases of firefighting and emergency response. They should also be trained in the technical aspects of their occupations to assist in the control of damage.
2 Arranges the organization into self-sufficient units. These units must have communication with each other. They must be able to take corrective action to control the various types of fires and emergencies.
3 Uses one central station such as DCC to receive reports from all DCRS stations. DCC evaluates and initiates those orders necessary for corrective action from a ship-wide point of view. This station also reports to and receives orders from the bridge (command control). These reports concern matters that affect the ship's stability and buoyancy, watertight integrity, and CBRN warfare defense measures.
4 Ensures that DCRS are assign tasks that are peculiar to a single department and under the direct supervision of an officer from that particular department.
5 Provides for relief of personnel engaged in difficult tasks, for battle messing, and for the transition from one condition of readiness to another. Develop procedures to ensure that all relief crews are informed of the overall situation.
6 Provides accurate and rapid communications between all firefighting and emergency response units. An overall coordination of effort and direction can then be readily accomplished.

7 Provides for a DCRS to assume the responsibilities of DCC in the event that DCC becomes non-functional.

Damage Control Central (DCC)

The emergency operations center for the DCA is in DCC during firefighting and various emergency situations. The Firefighting and Emergency Response teams are assigned to DCRS in the battle organization. BDS are located near the DCRS. Personnel assigned to the DCC are under the supervision of the DCA. These personnel perform the following tasks:

1 Receives and evaluates information from all DCRS.
2 Informs command control of conditions affecting the ship's stability and buoyancy and the ship's watertight integrity.
3 Provides direction to DCRS in regards to firefighting and controlling damage.
4 Controls watertight integrity, flooding, counter-flooding, and flood water removal.
5 Keeps command control informed about the progress of the following:
   a Combating damage, fire, and flooding.
   b Effects of CBRN attack.
   c Significant personnel casualties.
6 Evaluates the necessity of flooding the ammunition storage that are endangered by fire and recommend corrective action to the commanding officer. Directs DCRS to activate sprinkling to flood the necessary ammunition lockers when ordered by the commanding officer.
7 Displays charts and diagrams to show the ship's subdivisions, vital piping, and electrical systems.
8 Makes updates to DCC's casualty board to show sustained damage and the progress of the corrective actions. Ensures a simplified schematic is maintained on the bridge for visual reference by command control on the casualty data reported by DCC.

9 Makes updates to the stability board to show the liquid loading, the location of flooding boundaries, the effects on ship's stability and buoyancy caused by flooded rooms, and the corrective actions taken to improve the ship's stability and buoyancy. A liquid loading and flooding effects diagram is normally used for this purpose.

10 Prepares a list of BDS stations.

11 Prepares graphic displays to show corrective actions to firefighting and emergency response systems and ship's electrical systems.

12 Prepares deck plans to show all CBRN contaminated areas to include safe routes to decontamination stations and BDS.

13 Prepares a contamination prediction plot for radiological and nuclear attacks.

Damage Control Repair Stations (DCRS)

The DCRS leaders also known as the repair party leader (RPL) will take charge of activities in their area of responsibility after damage is sustained. They will keep DCC informed during all emergency situations. There are certain DCRS as shown in Table 2-1 and they may be subdivided to provide adequate protection for large areas. On aircraft carriers and large amphibious ships, the prescribed responsibilities may be the joint responsibility of two or more DCRS. When DCRS are subdivided, they are designated by the number of the parent repair station followed by a letter (such as IA, IB).

Outfitted with firefighting and emergency response equipment and the composition of the DCRS permit each repair station to handle the damage and casualties that occur within their assigned areas. Each ship must designate a DCRS as secondary DCC with a complete succession for command of firefighting and emergency response will be promulgated and posted in each DCRS. When succession of command is designated, the factors that are considered shall be the physical location of each DCRS, the seniority of each DCRS leaders, and the communication facilities available. The following general composition is considered necessary to ensure the effectiveness of the repair stations:

1 Damage Control Repair Station 1 (Main Deck).
2 Damage Control Repair Station 2 (Forward Ship).
3 Damage Control Repair Station 3 (Aft Ship).
4 Damage Control Repair Station 4 (Mid Ship).
5 Damage Control Repair Station 5 (Propulsion).
6 Damage Control Repair Station 6 (Ordnance).
7 Damage Control Repair Station 7 (Island Structure).
8 Damage Control Repair Station 8 (Electronics).

Damage Control Repair Station 1 (Main Deck)

An officer or senior enlisted crewmember from a deck division leads this repair station that is comprise of aviation, deck, and supply departments' personnel. On aircraft carriers, the hangar deck officer manages DCRS 1H that controls the Hangar Bay Firefighting and Emergency Response team. DCRS 1H is a subdivision of DCRS 1. A junior officer or senior enlisted crewmember is assigned as an assistant for the hangar bay. This team is responsible for the main deck of the ship and specifically responsible for the following:

1 Controls and extinguishes all fires.
2 Repairs damage in assigned areas.
3 Assists other DCRS and the crash and salvage team as required.

Table 2-1. Damage Control Repair Stations and Teams per Ship Class

Fire-fighting and Emergency Response Teams	Aircraft Carriers (CVN)	Large Amphibous Ships (LHA, LHD)	Surface Ships >225 FEET (CG, DDG, LPD, LSD)	Surface Ships >225 FEET (LCS)	Surface Ships <225 FEET (MCM)	Surface Ships <225 FEET (PC)
Damage Control Repair Station 1 (Main Deck)	X	X				
Damage Control Repair Station 2 (Forward Ship)	X	X	X	X	X	X
Damage Control Repair Station 3 (Aft Ship)	X	X	X	X		
Damage Control Repair Station 4 (Mid Ship)	X	X				
Damage Control Repair Station 5 (Propulsion)	X	X	X			
Damage Control Repair Station 6 (Ordnance)	X					
Damage Control Repair Station 7 (Island Structure)	X					
Damage Control Repair Stations 8 (Electronics)	X					
Damage Control Unit Locker (DCUL)	X	X	X			
Damage Control Unit Patrol Station (DCUPS)	X					
Damage Control Rescue and Assistance Reentry Locker (DCREL)	X	X	X	X	X	X
Aviation Fuel Repair Team and Crash and Salvage Team	X	X	X	X		
Ordnance Disposal Team	X	X				
At-Sea Fire Party (Flying Squad)/Rapid Response Team	X	X	X	X	X	X
In-port Emergency Teams	X	X	X	X	X	X

Damage Control Repair Station 2 (Forward Ship)

A properly trained officer or senior enlisted crewmember manages this repair station. This repair station assigns crewmembers from operations, deck, supply and weapons departments. Repair Station 2 will specifically respond to any emergency on the forward section of the ship as follows:

1. **This repair station is responsible for the forward section of the ship.** Clear the upper decks of wreckage that interferes with the operation of the battery, ship, or fire control stations. Clear wreckage that fouls the operations of the forward gun mount, anchor or sides of the ship.
2. Extinguishes all types of fires.
3. Maintains and makes emergency repairs to battle service systems. These systems include ammunition supply, ventilation supply, high- and low-pressure air lines, communications systems, electrical systems, and cooling water systems.
4. Provides emergency power to vital electrical equipment, using casualty power cables.
5. Assists other repair stations and the crash and salvage team as required.
6. Repairs damage above the waterline that could cause flooding in the event of further damage.
7. Repairs damage to structures, closures, or fittings that maintain watertight integrity. Shore, plug, weld, and caulk the bulkheads and decks; reset valves; and blank or plug lines through watertight subdivisions of the ship.
8. Maintains two status boards for accurate evaluation of underwater damage. The Stability Status Board (Flooding Effects Diagram) is a visual display of all flooding, flooding boundaries, corrective measures taken, and effects on ship's stability and buoyancy. The Liquid Load Status Board shows the current status of all fuel and water tanks and the soundings of each tank in feet and inches.
9. Maintains a graphic display board showing damage and action taken to correct disrupted or damaged systems.

10 Maintains a casualty board for visual display of structural damage
11 Administers first aid to injured personnel and then transport them to without seriously reducing the firefighting and emergency response capabilities of the repair station.
12 Maintains deck plans showing locations of CBRN contamination and safe routes to personnel decontamination and BDS.
13 Detects, identifies, and measures dose and dose-rate intensities from radiological involvement.
14 Surveys potential radiological contaminated areas and personnel; decontaminate areas and personnel that received radiological contamination.
15 Obtains biological agent samples for identification of biological agents.
16 Identifies any potential chemical agents used during an attack.
17 Conducts decontamination of all areas and personnel affected by biological or chemical attack.
18 Evaluates and reports the extent of damage sustained in their areas.

Damage Control Repair Station 3 (Aft Ship)

Like DCRS 2, a junior officer or senior enlisted crewmember manages this repair station. This repair station assigns crewmembers from operations, deck, supply, and weapons departments and a few personnel from engineering departments. This repair station will carry out the same responsibilities as DCRS 2 but specifically respond to any emergency on the back section of the ship.

Damage Control Repair Station 4 (Mid Ship)

Similar to both DCRS 2 and 3, a junior officer or senior enlisted crewmember manages this DCRS. This repair station assigns crewmembers from operations, deck, supply, and weapons departments and a few personnel from engineering departments. This team will carry out the same responsibilities as DCRS 2 and 3. This repair station will respond to any emergency occurring in the middle section of the ship.

Damage Control Repair Station 5 (Propulsion)

An engineering department officer will manage this DCRS. This firefighting and emergency response team is made up of various engineering occupations. In the assignment of personnel emphasis should be placed on their resident expertise and lay-out of the engine, auxiliary and electrical plant rooms. This team will specifically maintain and respond to emergencies occurring to the ship's propulsion, auxiliary and electrical plant rooms as follows:

1 Make repairs or isolate damage to main propulsion, auxiliary and electrical plant machinery.
2 Extinguishes all fires in engine, auxiliary, and electrical plants.
3 Operates, repairs, or isolates vital systems. Modify the methods of segregating vital systems when necessary.
4 Repairs to any damage that could potentially cause flooding in the event of further damage.
5 Assists in the operation and repair of the steering control systems.
6 Assists in the maintenance and repair of communications systems.
7 Assists other repair stations and the crash and salvage team when required.

Damage Control Repair Station 6 (Ordnance)

An officer or senior enlisted crewmember manages this DCRS. The firefighting and emergency response team is comprised of personnel from weapons and combat systems departments. This team will respond to emergencies occurring to weapon and combat systems and may be divided into subgroups 6A for forward and 6B for aft. On surface combatant ships not using DCRS 6, ordnance emergencies and repairs are conducted by teams within the combat system casualty control organization with assistance from other DCRS. This team shall be responsible for the following:

1 Make emergency repairs to all ordnance installations.
2 Operates the magazine firefighting sprinkler systems and other ordnance systems with permission from the commanding officer.
3 Assists other DCRS in extinguishing fires in the vicinity of magazines.
4 Assists other DCRS in making hull damage repairs.
5 Station repair party control at the forward magazine sprinkler control station.
6 Maintain communications with Weapons Control, DCC, and its own detached units.
7 Isolate affected magazines from other magazines in the same group.
8 Notifies DCC during the activation any sprinkling/flooding firefighting system for each magazine. Only the commanding officer may authorize the flooding of Magazines.

Damage Control Repair Station 7 (Island Structure)

This DCRS is only required on aircraft carriers and may be necessary on large deck amphibious ships. An appropriately trained officer or senior enlisted crewmember manages this repair station. The firefighting and emergency response team is made up of personnel from the air and engineering departments but additional personnel can be augmented from other departments when necessary. This DCRS shall respond to emergencies in the island structure or tower of the aircraft carrier. This repair station shall carry out the following actions:

1 Controls and extinguishes all fires.
2 Repairs from battle damage in assigned areas.
3 Assists DCRS 1 and other DCRS as required.

Damage Control Repair Station 8 (Electronics)

This repair station is managed by an officer or senior enlisted crewmember from the operations, weapons, and combat systems department. This electronic casualty control team is comprised of personnel from electronics and electrical occupations. This team shall respond to all emergencies on ships with highly complex electronic weapons systems such as missile ships and aircraft carriers. This team will meet its responsibilities as follows:

1. Repairs radar, radio, countermeasures, and all associated electronics equipment.
2. Repairs fire control equipment.
3. Repairs sonar equipment.
4. Extinguishes minor electrical fires.
5. Assists other DCRS when required.

Damage Control Unit Locker (DCUL)

In addition to DCRS, various ships utilize DCUL located in passageways throughout the ship. These DCULs are used as a rapid dispersal of personnel and equipment to extend a DCRS's investigation and initial response capabilities within the assigned area of responsibility. An officer or senior enlisted crewmember manages this small team comprised of personnel from various occupations. This team will meet its responsibilities as follows:

1. Provides a rapid means to thoroughly investigate its assigned area.
2. Provides an immediate response to initiate firefighting actions, setting fire and flooding boundaries.
3. Requests assistance from its parent DCRS if the situation becomes overwhelming.
4. Ensures that watertight integrity is not jeopardized and that classified fittings are maintained in the state of closure required by the material condition in effect.

Damage Control Unit Patrol Station (DCUPS)

Aircraft carriers and various large amphibious ships also have manned DCUPS for rapid response to difficult-access areas with high potential for damage. This locker is smaller than a DCRS but larger than a DCUL. This team is led by an officer or senior enlisted crewmember and it is made up of personnel from various occupations. This team has similar responsibilities as a DCUL except it has the capability to respond to fire and flooding emergencies.

Damage Control Rescue and Assistance Reentry Locker (DCREL)

Ships such as aircraft carriers, destroyers, patrol crafts, mine countermeasure ships, amphibious ships, and new construction ships are required to have an unmanned compartment easily accessible to the weather decks. Depending on the class of ship, DCREL can function as a DCRS but it is primarily used to render rescue and assistance to other ships and crafts in danger. In addition to rescue and assistance functions, the DCREL can provide firefighting and flooding capabilities to the ship when required. The fire marshal acts as the RPL whenever the need for a Rescue and Assistance emergency. This team is primarily made up of fire protection engineers, hull maintenance technicians, machinery repairman, and other occupations.

Aviation Fuel Repair Team and Crash and Salvage Team

These teams are required on all air-capable ships such as aircraft carriers and other ships that are equipped for manned flight operations. On aircraft carriers and large amphibious ships, an officer or senior enlisted crewmember from the air department will manage this team comprised of personnel from the air department. On other ships that are equipped for manned helicopter operations, appropriate deck, engineering department, and fire protection engineering personnel are assigned.

An appropriately trained officer or senior enlisted crewmember is trained to lead this team. This team is responsible for the following:

1 Operates, maintains, and make repairs to all aviation fuel systems.
2 Controls and extinguishes fires in aircrafts, on the flight deck, and hangar bay.
3 Promptly rescues pilots after crash of aircrafts.
4 Conducts flight deck repairs and removes wreckage to restore the flight deck for normal operations.

Ordnance Disposal Team

The ordnance disposal team is comprised of specially trained personnel deployed aboard ships as required. The team is organized unit within the ship's weapons department. The ordnance disposal team normally operates under the direction of the ship's weapons officer. This team stabilizes unexploded ordnance or removes as necessary.

Battle Dressing Stations (BDS)

The BDS functions as a triage and first aid station. Most ships have a minimum of two BDS equipped for emergency treating personnel battle casualties. However, many smaller ships, such as minesweepers, have only one such station. Those ships having two or more stations should have the stations well separated from each other. Each station must be accessible to the stretcher bearers from DCRS within the vicinity. Medical department personnel are assigned by the senior member of that department. Additional crewmembers trained in CPR and first-aid is used to supplement each station. First-aid kits or boxes are available at DCRS and BDS. The medical department furnishes the material for these first-aid kits and boxes.

At-Sea Fire Party (Flying Squad)

Most surface ships have organized a special fast-response firefighting and emergency response team known as the at-sea fire party. This team is sometimes called the "Flying Squad" because its fast-response capabilities. The ship's fire marshal is in charge of this team. This team is primarily comprised of fire protection engineer, hull maintenance technician, machinery repairman and other highly trained crewmembers. This team is used to respond to emergencies without disrupting normal operations of the ship while at-sea and in-port during normal working hours. A key component of the Flying Squad is the Rapid Response Team consisting of four members who reports directly to the scene as initial responders until relieved by the Flying Squad. The Flying Squad will carry out the following responsibilities:

1. Respond immediately to fire alarms when the ship's DCRS are not manned.
2. Extinguish various fires without disrupting other ship operations.
3. Control fires until ongoing sensitive critical evolutions can be secured and GQ or battle stations can be manned and ready.
4. Respond to flooding and make necessary repairs.
5. Conduct CBRN warfare defense operations during working hours while in-port.
6. Isolate and remove toxic gases and hazardous materials.

Inport Emergency Teams (IET)

The IET will function as a DCRS while the ship is in-port during after-hours, weekends and holidays. This team is made up of various personnel assigned to a duty section which covers a 24-hour shift. The engineering duty officer (EDO) along with assistance from the duty fire marshal is in control of this team. This team shall carryout the following responsibilities:

1. Extinguish all fires.
2. Make emergency repairs on damage from collision or sabotage.

3 Respond to flooding and make necessary repairs.
4 Isolate and remove toxic gases and hazardous materials
5 Conduct CBRN warfare defense operations.
6 Assist the Flying Squad during normal in-port working hours.

Firefighting and Emergency Response Team Positions

No two emergency situations are identical; therefore, the corrective action taken may vary. The firefighting and emergency response must be coordinated with other elements of the ship's organization to achieve these goals. Each department must assign individuals to specific firefighting and emergency response positions. The responsibilities of each firefighting and emergency response team member will normally remain the same but may be required to assume other roles. For example, during fire emergencies the team is called a hose teams consisting of a nozzleman, hoseman, and plugman. Whenever there is a flooding emergency the hose team members will become part of teams used to minimize the flooding as shown in Table 2-2. These teams are called the pipe-patching team, plugging team, shoring team, and dewatering team. The assignments of the firefighting and emergency response team will change to accommodate the emergency situation at hand.

Table 2-2. DCRS Functions and Manning Requirements

Fire-fighting and Emergency Response Team Position	When Required (Condition 1 - General Quarters and II DC)	At-Sea Fire Part (Flying Squad)	In-port Emergency Team (IET)	Rescue and Assistence Team (R&A)
Officer in Charge (OIC)	N/A	N/A	N/A	Fire, Flooding
File Marshal	Fire, Flooding	Fire, Flooding	Fire, Flooding	N/A
Damage Control Repair Station Leader or Repar Party Leader (RPL)	Fire, Flooding	N/A	N/A	N/A
On Scene Leader (OSL)	Fire, Flooding	Fire, Flooding	Fire, Flooding	Fire, Flooding
Team Leader	Fire, Flooding	Fire, Flooding	Fire, Flooding	Fire, Flooding
Nozzleman*	Fire	Fire	Fire	Fire
Hoseman*	Fire	Fire	Fire	Fire
Plugman*	Fire	Fire	Fire	Fire
Electrician	Fire, Flooding	Fire, Flooding	Fire, Flooding	Fire, Flooding
Investigator	Fire, Flooding	Fire, Flooding	Fire, Flooding	As Required
Boundaryman	Fire, Flooding	Fire, Flooding	Fire, Flooding	Fire, Flooding
Smoke Controlman*	Fire	Fire	Fire	Fire
AFFF Station Operator (As Required for AFFF System)	Fire	Fire	Fire	N/A
SCBA Coordinator	Fire	File	Fire	As Required
Communicator/ Phone Talker/ Plotter/Messenger	Fire, Flooding	Fire, Flooding	Fire, Flooding	N/A
Pipe Patching	Flooding	Flooding	Flooding	Flooding
Shoring	Flooding	Flooding	Flooding	Flooding
Hull Patching/ Plugging	Flooding	Flooding	Flooding	Flooding

Fire-fighting and Emergency Response Team Position	When Required (Condition 1 - General Quarters and II DC)	At-Sea Fire Part (Flying Squad)	In-port Emergency Team (IET)	Rescue and Assistence Team (R&A)
Dewatering	Flooding	Flooding	Flooding	Flooding
Stretcher-Bearer/First-Aid (As Required for Personnel Injury)	Personnel Casualty	N/A	First Aid	First Aid

*In regards to the Flying Squad, IET, R&A and various small ships, the Nozzleman, Hoseman Plugman may be required to serve as members of the Pipe Patching, Shoring, Hull Patching/P lugging, and Dewatering teams during flooding emergencies.

Fire Marshal

When there is an emergency as prescribed in the general emergency bill, the fire marshal shall proceed directly to the scene of the emergency to direct efforts of the Rapid Response Team. If the emergency is beyond their capabilities, the team shall isolate the casualty to prevent spreading and the fire marshal shall turn actions over to the on-scene leader (OSL) and assume other duties as directed. The fire marshal must assume a "big picture" role upon being relieved by the OSL providing particular attention to the potential for vertical fire spread or other spreading of damage. The fire marshal will make recommendations for additional Condition II Damage Control or General Quarters as required by the magnitude of the casualty. The fire marshal duties may include:

1 Becomes the DCRS leader for the Flying Squad.
2 Ensures communications are established and maintained.
3 Ensures boundaries are established.
4 Provides direct logistic by supplying equipment to the Flying Squad.
5 Assists the DCA in DCC or becomes a DCRS leader when Condition II Damage Control or General Quarters is set.
6 Conducts post-fire atmospheric testing.

Damage Control Repair Station Leader

The DCRS leader or RPL coordinates all firefighting and emergency response actions in their assigned area of the ship. The RPL is an officer or senior enlisted crewmember and they manage all personnel and material assigned to that repair station. They will report to their assigned DCRS and establish necessary lines of communication with DCC. They shall be totally familiar with the repair stations area of responsibility, its equipment, and ensure available assets are properly utilized to control and extinguish all fires and handle all emergencies as prescribed by the general emergency bill. The Repair Party Leader shall:

1. Dispatches investigators to discover damage, establish fire and smoke boundaries, flooding boundaries, and toxic gas boundaries.
2. Establishes communications with DCC, other DCRS, OSL and investigators.
3. Dispatches firefighting and emergency response teams to applicable casualties.
4. Manages the Boundarymen via the investigators.
5. Maintains control of operations within their area of responsibility and coordinate actions with the DCA.
6. Decides appropriate actions based on reports from investigators and the OSL.
7. Tracks self-contained breathing apparatus (SCBA) activation times of all firefighting and emergency response team members.
8. Provides relief teams for the firefighting and emergency response teams.
9. Ensures all boundaries are established and maintained, with particular attention given to the potential for vertical fire spread.
10. Ensures mechanical, electrical, and hull isolation is established to the degree required by the situation.
11. Provides logistic support to the scene of the emergency.
12. Ensures an accurate diagram plot is maintained and necessary status reports are made.

On-Scene Leader (OSL)

The OSL is the person in control at the scene of the emergency. This position is usually held by a fire protection engineer or any highly knowledgeable crewmember that is qualified in this position. When an emergency is called away, the OSL will proceed immediately to the scene after dressing in proper protection, equipped to assume control of the firefighting and emergency response team. Once the OSL reports to the scene of the casualty, the OSL will assume control of emergency actions. The OSL will receive reports from team leader and investigators and pass on those reports to the DCRS leader.

The OSL determines what resources are needed for a firefighting and emergency response team, including the need for a separate individual as team leader and a backup hose team. The scene leader shall:

1 Wears a SCBA equipped with voice amplifier.
2 Immediately assess extent of fire and emergency situation.
3 Determines the require equipment for the emergency situation.
4 Determines the firefighting agent to be used.
5 Determines the method and direction of attack for fires and flooding emergencies.
6 Be positioned to best control the firefighting and emergency response team.
7 Determines to secure electrical power to the effected space.
8 Determines the number of hose team members needed.
9 Tracks SCBA activation times for hose team members.
10 Establishes communications as required using best means.
11 Determines the protective clothing requirement for the firefighting and emergency response team, based on assessment of conditions found.
12 Knows how to operate all installed firefighting systems, portable fire extinguishers, and firefighting equipment.

Team Leader

The attack team leader is in direct control of the firefighting and emergency response team and makes periodic reports to the OSL. During a fire the team leader is responsible for navigating the hose team(s) throughout the compartment for the attack on the fire, directing the in-space method of extinguishment, and maintaining situational awareness for the safety of the hose team. The team leader must be able to do the following:

1. Ensures the team is dressed in proper protective clothing with SCBA.
2. Ensures fire hose and firefighting agent is tested and operational.
3. Knows the layout of the affected space, and be aware of any hazards.
4. Knows various firefighting methods and techniques.
5. Directs hose team around hazards to reach the source of the fire.
6. Directs the efforts of hose team to extinguish or overhaul a fire.
7. Uses the thermal imager to locate fires and hot spots.
8. Directs the nozzleman in selecting the spray pattern to be used.
9. Directs the rotation of hose team personnel to mitigate fatigue and heat stress.
10. Once the fire is extinguished, use thermal imager to overhaul the effective space.
11. Determines proper action for containing and stopping flooding.
12. Makes periodic reports of conditions to the OSL.
13. Knows how to operate all installed firefighting systems, portable fire extinguishers, and firefighting equipment.

Investigator

Investigators are assigned to the firefighting and emergency response teams to ensure that no further damage occurs outside the boundaries of the existing fire or flooding casualty. Investigators operate in pairs, travel assigned routes, and report conditions to the DCRS. The investigator will ensure that the boundaries around the casualty are maintained and that further damage is not occurring. It may be necessary for them to access locked spaces to check for damage and they will carry tools to open these spaces. A person assigned to this position must have an in-depth knowledge of the ship's layout and the systems that are in their assigned area. When conducting an investigation, they shall be required to do the following:

1 Properly dresses in protective clothing along with a SCBA.
2 Establishes communications with OSL and RPL.
3 Monitors boundaries on existing casualty to prevent spread of damage.
4 Investigates thoroughly around casualty area.
5 Investigates cautiously while looking for damage.
6 Reports findings immediately to the OSL and RPL.
7 Size-ups the damage when discovered, isolates damage and establishes boundaries.
8 Repeats investigation of casualties.
9 Looks for hidden damage during repeated investigation.
10 Knows how to operate all installed firefighting systems, portable fire extinguishers, and firefighting equipment.

Nozzleman

The nozzleman is the person that controls and operates the firefighting hose nozzle. They will be safely directed by the team leader through the compartment to extinguish the fire. They will be required to wear proper protective clothing and breathing apparatus. Once the fire is extinguished and the nozzleman becomes the reflash watch and may be involved in the overhaul process of the fire. The primary responsibility of the nozzleman as follows:

1 Directs the firefighting agent towards the fire.
2 Knows the proper pattern and flow rate of the firefighting nozzle.

3 Knows various firefighting methods and techniques.
4 Knows how to operate all installed firefighting systems, portable fire extinguishers, and firefighting equipment.
5 Acts as the team leader when required.
6 Wears and activates SCBA.

Hoseman

The hoseman is required to help support the nozzlemen and tend the hose. The number of hoseman will depend on the size of hose, the location of the fire, and how many turns or bends it takes to reach the fire. The number of hosemen will be determined by the OSL. The minimum hosemen needed for a 1-½ inch or 1-¾ inch firehose is three personnel consisting of two hoseman and one nozzleman. The minimum hosemen needed for a 2-½ inch firehose is five personnel consisting of four hoseman and one nozzleman. They will run the firehose from a fireplug to the location of fire. They will ensure to keep the hose from getting fouled or entangled while fighting the fire. The nozzleman will be able to handle the hose and nozzle better when the hoseman keep the weight and tension of the hose off the nozzleman. Other responsibilities of the hoseman shall include the following:

1 Handles the AFFF hose and freshwater hose when required.
2 Knows how to operate all installed firefighting systems, portable fire extinguishers, and firefighting equipment.
3 Relays spoken updates, requests and directions between the OSL, team leader, and nozzleman.
4 Maintains a position to support movement of hose around choke points (bends or corners).
5 When directed assumes the duties of the nozzleman and reflash watch.
6 Acts as the access person to open up the door to enter the effective area and use entry forcible tools to gain access and clear routes.

7 Assists with overhaul and clean up after the fire has been extinguished.

Plugman

The last hoseman will act as the plugman. They will ensure to open the fireplug before the hose team enters the effective space. The will be responsible for the following:

1 Ensures all firehoses are laid out and connected to firefighting nozzle and fireplug valve. When directed opens the fireplug valve to charge the firefighting hoses.
2 Ensures AFFF hose is laid out. When directed activates the AFFF hose reel.
3 Ensures freshwater hose is laid out. When directed activates the freshwater hose reel.
4 Knows how to operate all installed firefighting systems, portable fire extinguishers, and firefighting equipment.
5 Monitors the hose for loss of pressure or a hose rupture. In the event of loss of pressure, makes a report to the OSL and takes proper actions to restore firefighting agent.
6 In the event of a ruptured fire hose, secure the fireplug, replace the ruptured section of hose, and reactivate the hose as quickly as possible.
7 Assists in rigging a jumper hose or setting up portable pumps such as the P-100.
8 Connects firefighting foam suction tube and place suction tube inside 5-gallon can of firefighting foam when required.
9 Monitors the level of firefighting foam inside 5 gallon cans and replace cans.

Electrician

In the event of fire or flooding, the repair station electrician will be directed to immediately secure electrical power to all shipboard compartments that are affected by the casualty. A repair station electrician may be assigned to the Flying Squad's Rapid Response Team and they will immediately report to the casualty area and follow instructions given by the fire marshal. When assigned to a DCRS, the electrician will update the OSL and DCRS leader when the electrical power is secured. The electrician is responsible for the operating various electrical powered firefighting and emergency response equipment. The electrician may also be responsible for the following:

1. Investigates for electrical damage once the casualty is corrected.
2. Rigs casualty power (emergency power) to vital shipboard systems.
3. Conducts repairs to vital electrical systems as soon as possible.
4. Conducts repairs to non-vital electrical systems once vital systems have been restored.

Boundaryman

Boundarymen proceed directly to the scene when a fire or flooding casualty is called away to commence establishing primary and secondary boundaries as directed by the fire marshal or DCRS leader. On various ships organization the boundarymen can be a part of the rapid response team. Their initial actions are to secure all doors, hatches, and openings in the boundary of the affected area. The ideal is to close the casualty in like a box by establishing boundaries on all six sides of the affected space. During fires, the most important boundary that must be established is the above boundary. The below and side boundaries must be immediately set during a flooding situation. The boundarymen shall be responsible for the following:

1. During a fire they will remove or relocate combustibles from the boundaries.

2 If they notice that the deck (floor) or bulkhead (wall) begins to heat up, they will cool the boundaries with fire hoses, buckets of water or by any means necessary.
3 They are monitored by the investigators and they will make continuous reports to the DCRS leader or investigators.
4 In the case of flooding, the first group of watertight transverse bulkheads forward and aft of the flooded compartment will be used as the flooding boundaries.
5 Thoroughly monitors closures, walls, seams and joints of the compartment. They will immediately make reports if any closures, seams, or joints begin to leak or if the bulkhead starts panting or overhead starts sagging (wall or ceiling starts buckling).

Smoke Controlman

The smoke controlman are used in conjunction with the boundarymen during a fire casualty. They will ensure that smoke boundaries are established around the affective area. This area is known as the smoke control zone and they will use the box method to help establish this zone. They can also function as the desmoking team when directed by the DCRS leader. They will be responsible for the following.

1 Secures ventilation fans or closes vent dampers in the ship's ventilation system.
2 Hangs smoke curtains on open accesses that do not have doors.
3 Rigs portable fans when implementing active desmoking at the access area to the affected compartment.
4 Rigs portable fans and blowers when implementing post-fire desmoking.

AFFF Station Operator

The AFFF station operator is used during a fire that involves fuels in the ship's engine room, auxiliary room, fuel pump room, flight deck and hangar bay. During a fire in any of the above spaces, the AFFF operator will immediately report to their assigned AFFF station. During flight operations on various ships an AFFF operator is required to report to the AFFF station. They will be required to perform the following tasks:

1 Ensures that there is a constant supply of AFFF
2 Ensures that operating pressure is maintained
3 Establishes communications with DCRS leader, DCA, and other AFFF stations.
4 Be knowledgeable in the operation of the ship's AFFF system,
5 Refills the AFFF tank as necessary.

SCBA Coordinator

When established by the DCRS leader, the SCBA coordinator will manage breathing air for the firefighting and emergency response team at a designated SCBA recharging station or SCBA change-out station. Each ship shall prepare procedures for breathing air management during firefighting. The extent to which breathing air management functions are implemented depends on the casualty. They may be assigned other functions in addition to the following responsibilities:

1 Assists personnel in obtaining the correct size air cylinder for their assigned duties and assist personnel in donning their SCBA.
2 Assists personnel in recharging or removing their SCBAs or changing-out their cylinder.
3 Ensures that sufficient spare SCBA air cylinders are supplied to the SCBA Change-out Station(s).
4 Ensures that all depleted SCBA air cylinders are recharged.
5 Reports SCBA start times.
6 Assists DCRS leader with tracking SCBA start and stop times.

Communicator

The communicator may function as a phone talker or messenger depending on the ship's firefighting and emergency response organizational make-up and communications equipment configuration. They must be able to operate the designated DC communications circuits. They may use a computer system, wireless free communications radios (WIFCOM), ships telephone system, sound-powered phones, or written messages. If they are assigned as a phone talker, they will report to DCC, the DCRS, or other stations such as the bridge. They will establish communications between the assigned station and other stations within the firefighting and emergency response organization. They will receive messages from other phone talkers and relay them to the stations supervisor.

The use of a messenger is the last means of communications used whenever the above systems fail. If a messenger is needed then they will be responsible for carrying messages between the scene of the casualty, DCRS, DCC and OOD. They must be thoroughly familiar with the lay-out of the ship and know how to get from one place to another. They must stay near their assigned area at all times except when they are taking messages back and forth. When carrying a message from one point to another, they must do it quickly because there maybe another message ready to be routed.

Stretcher-Bearer

A stretcher-bearer is assigned to the BDS to assist medical department personnel. They are required to transport first-aid kits or boxes to a clear area near the scene of the casualty and administered first aid, as required and transport injured personnel carefully through constrained passages within and around the ship to the designated triage area. Each ship will vary but the minimum number of stretcher-bearers will be four per each triage and first-aid stations. The use of a BDS station will be determined by the tiered conditions of readiness. For example, the DCA can order a BDS station to be established during Condition II DC. This will allow a quick response to aid any injured personnel.

This chapter has covered the organization of the firefighting and emergency response program aboard ship and the responsibilities of each DCRS and its team. This program is broken down into administrative organization and battle organization. The key to success is readiness that these two organizations provide. The administrative is needed to ensure that equipment is immediately ready to use during emergencies while the battle organization is properly trained to respond to all fires and various emergencies that could occur from battle or routine operations. The conditions of readiness give a ship the flexibility to remain in normal operations during certain emergencies. The entire ship's company must be trained and understand the necessity for maintaining the highest degree of efficiency in firefighting and emergency response. Each personnel that are assigned to DCRS, the Flying Squad, or an IET must be properly trained and ready to step up to the next role to ensure the survival of the ship during emergencies. The next chapter will cover the fundamentals of shipboard fire protection which is significant for the firefighting teams to understand when responding to any class of fires.

CHAPTER 3

Fundamentals of Shipboard Fire Protection

Fires are potential hazards aboard ships and submarines that will also be an endless threat that personnel must take all possible measures to prevent from occurring. If a fire does occur, personnel must immediately report the fire and then it must be rapidly extinguished. Often a fire may occur from actions caused by enemy attacks or coincidental from storms or accidents. If not immediately extinguished, a fire can cause more damage to the ship than the initial damage that caused it or the worst case scenario would be the loss of the ship as shown in figure 3-1. A fire protection engineer must identify each class of fires, apply the preferred extinguishing methods, and operate all firefighting systems and equipment. Also, they must conduct maintenance and repairs to all the firefighting systems and equipment which will effectively contribute to the overall safety of the ship.

Figure 3-1. ***USS Belknap*** **(CG 26) 1975**

If a fire continues to burn without confining or extinguishing it, the fire can spread rapidly; therefore, speed is very important factor in controlling and extinguishing fires. A small trash can fire will easily spread to other combustibles items and become a blazing inferno throughout several compartments affecting the entire ship. The cost of damage can arrange from a few thousands of dollars to millions of dollars; therefore, the ship's firefighting team must immediately get to the scene and begin firefighting evolutions utilizing all available systems and equipment. Any delays will allow the fire to rapidly spread making it more difficult to extinguish the fire. So it is critical that firefighting equipment and systems are operational and available to the firefighting team. The team must be knowledgeable in the purpose and use of each equipment and system. The firefighting equipment will allow them to extinguish the fire so the ship can continue the mission. This chapter covers the fundamentals of firefighting to include classification of fires, the effects of fire, the fundamentals of extinguishing fires, and the firefighting agents used to extinguish the fire. This chapter will also discuss firefighting extinguishers, firefighting systems, personal protection gear, and portable equipment available for the firefighting team.

Start of a Fire

Matter exists in three states known as solid, liquid, or vapor (gas). The atoms or molecules of solid matter are arranged closely together, whereas the liquid molecules are loosely arranged. The molecules of vapor matter are not arranged together moving freely. In order for a substance to burn, oxygen molecules must surround the substance. The solid and liquid molecules are too closely arranged to be surrounded by oxygen molecules, whereas only loose molecules of vapors can burn. When a solid or liquid substance is heated, the substance molecules begins to move more rapid and if enough heat is applied, some of the molecules break away from the surface to form a vapor just above the surface of the substance. These freely moving vapor molecules are now mixed with the oxygen molecules. If the substance is continuously heated, the vapor's temperature will rise to its ignition temperature and the substance's vapor will ignite or combust when there is enough oxygen present. Ignition or combustion involves the rapid oxidation of millions of vapor molecules. The molecules oxidize by breaking apart into individual atoms and recombining with oxygen into new molecules. It is during the breaking and recombining process that energy is released as heat and light which will turn into a fire.

Fire Components

In order for a fire to exist three components of combustible material, adequately high temperature, and a continuous supply of oxygen are required. These three components are easily referred to as heat, fuel, oxygen. Together these three components are known as the "fire triangle" (Figure 3-2). The fire triangle is appropriate for describing the requirements for surface glowing or smoldering, but does not completely describe the requirements for flaming combustion. A fourth component known as uninhibited chemical chain reaction is required for flames to exist. When uninhibited chemical chain reaction is combined with the other components of the fire triangle, it becomes a fire tetrahedron.

Fires are generally controlled and extinguished by eliminating one side of the fire tetrahedron. If the fuel, heat, or oxygen is removed, the uninhibited chemical chain reaction will be interrupted and the fire will be extinguished.

Figure 3-2. Fire Triangle

Heat

A fire is sometimes referred to as combustion because the process of combustion involves the rapid oxidation of millions of vapor molecules. The molecules oxidize by breaking apart into individual atoms and recombining with oxygen into new molecules. It is during the breaking and recombining process that energy is released as heat and light. Combustion involves rapid oxidation which is the chemical reaction by which oxygen combines chemically with the elements of the burning substance. Even when oxidation progresses slowly such as the rusting process in a piece of metal, a small amount of heat is generated in the process but, this heat will dissipate before there is any perceivable rise in the temperature of the metal being oxidized. Slow oxidation in certain types of materials can turn into fast oxidation if heat is not dissipated. A fire will occur from a unique process known as spontaneous combustion; therefore, materials that are identified to spontaneous combustion are stowed in a confined space where the heat can be dissipated rapidly. An example of materials that are subjected to spontaneous combust may be rags or papers that are soaked with animal fat, vegetable fats, paints, oils or solvents. This is why ships have designated storage areas or containers for these materials.

For combustible fuels or substances to ignite, it must have an ignition source and produce enough heat to burn. The lowest temperature at which a combustible material gives off enough vapors that will burn when a flame or spark is applied is known as the flash point. The temperature at which a substance will continue to burn after it has been ignited is known as the fire point. The fire point is normally a few degrees higher than the flash point. The auto-ignition or self-ignition point is the lowest temperature to which combustible materials must be heated to give off enough vapors that will burn without the application of a spark or flame. The auto-ignition point is the temperature at which the phenomenal process known as spontaneous combustion occurs. The auto-ignition point will normally be much higher temperature than the fire point. The flammable range or the explosive range is known as the range between the smallest and the largest amounts of vapor in a given quantity of air that will burn or explode when ignited. For example, a combustible material has a flammable or explosive range of 1 to 12 percent. In other words, either a fire or an explosion can occur if the atmosphere contains more than 1 percent but less than 12 percent of the vapor of this substance. Overall, the percentages referred to in connection with flammable or explosive ranges are percentages by volume.

Fuel

Fuels take on a wide variety of characteristics as the type of fuel will determine the class of fire and the extinguishing method. Fuels are found in forms of solids, liquid, or gases (vapors). The most recognizable solid fuels are wood, plastic, rubber, paper and cloth. These solid fuels are found throughout the ship as insulation, canvas, dunnage, packing, electrical cabling, wiping rags and mattresses. The flammable liquids that are most commonly found aboard ships are JP-5 fuel, Navy Distillate (F-76) fuel, diesel fuel (DFM), gasoline (MOGAS), JP-8 fuel, lubricating oil, hydraulic fluid, and oil base paints and their solvents. These liquids have loosely arranged molecules and can release vapors over a wide temperature range. The more the liquid is heated the more the vapor is released becoming easily ignited from a simple spark. Flammable gases that are commonly found on board ships include acetylene, MAPP gas, propane and hydrogen. Flammable gases are already in the vapor state and when added to the proper ratio with oxygen a sufficient heat source will cause ignition. The list of fuels is extensive but these are only a few examples.

Oxygen (O_2)

The oxygen (O_2) side of the fire triangle refers to the oxygen content of the surrounding atmosphere. A minimum concentration of 15 percent oxygen in the atmosphere is only required to support flaming combustion, whereas smoldering combustion can take place in an atmosphere containing only 3 percent oxygen. The atmosphere normally contains about 21 percent oxygen, 78 percent nitrogen, and 1 percent of argon and other gases. If an enclosed fire compartment is abandoned, closing all entrances and openings can reduce the supply of fresh oxygen and limit the burning rate of a fire.

Oxygen can be produce by material by the oxidizing process. These oxidizing materials release oxygen when it is heated and may react readily with other materials. Such substances include the hypochlorites, chlorates, perchlorates, nitrates, chromates, oxides and peroxides. All these substances supply enough oxygen to support combustion. Oxygen is released when the materials break down, as in a fire. For this reason, burning materials with their own oxidizers cannot be extinguished. Instead, large amounts of water are needed to cool surroundings while permitting a controlled burnout. Oxidizers are hazardous materials and must be stored in approved locations.

Fire Classifications

As mention before fires are classified according to the nature of the fuels involved. Identifying the classification of any fire is vital as it determines the manner in which agent and method utilized to extinguish the fire. Matters can become worse if the wrong type of firefighting agent is used on the wrong type of fuel. So it is very important to understand what is burning to classify the fire as ALPHA, BRAVO, CHARLIE, or DELTA.

Class ALPHA Fire

Class ALPHA (A) fires occur in common combustible materials such as wood, textile, cloth, paper, upholstery, fibrous products, and similar materials. Class A fires are mainly extinguished with water in high or low velocity fog or solid streams patterns, but if the fire is deep-seated, AFFF is more effective than sea water. In this case, AFFF can be used as a wetting agent to immediately penetrate and extinguish the deep-seated fire. Class A fires must be overhauled immediately after it is extinguished because it leaves embers or ashes that may potentially restart the fire.

Class BRAVO Fire

Class BRAVO (B) fires occur in flammable liquids such as gasoline, jet fuels, diesel oil, fuel oil, paints, thinners, solvents, lubricating oils, and greases. These fires are primarily extinguished with AFFF, halon 1211, halon 1301, heptafluoropropane (HFP), water mist or potassium bicarbonate ($KHCO_3$). Class B fires also involve flammable gases such as acetylene which should never be extinguished unless there is reasonable certainty that the flow of gas can be secured. Securing the fuel source is the single most important step in controlling a gas and fuel fire. The circumstances of the fire will determine the firefighting agent that is used.

Class CHARLIE Fire

Class CHARLIE (C) fires occur in electrical equipment and these energized electrical fires are attacked at specified distances using nonconductive agents such as carbon dioxide (CO_2), halon 1211, PKP or water fog spray but the most effective tactic is to de-energize and handle the fire as a Class A fire. When fires are not deep-seated, the preferred agents are CO_2 or halon 1211 because it does leave any residue. The last method of choice is to use water. The firefighter must use a water fog pattern and must cautious to never use a solid stream because water because conductive when use in a direct stream.

Class DELTA Fire

Class DELTA (D) fires occur in combustible materials or metals such as magnesium, titanium, sodium, and lithium. Metal fires on board ships are commonly associated with aircraft wheel structures. Special techniques are used to extinguish these types of fire such as applying large amounts of water on the burning material to cool it down below its ignition temperature, covering the burning material with large amounts of dry sand or heaving the burning material into the ocean. A small explosion may occur once water is applied to the burning material and the firefighter must take precautions by applying water from a safe distance or from behind shelter.

The Effects of Fire

Fire produces chemical reactions such as flames, heat, and smoke and each of these can cause serious injuries or death. Burning materials also produce various gases and other combustion products. Oxygen levels will be reduced by the gases and combustion products making it difficult for breathing. Because these chemical reactions from fires can cause death, it is extremely important to firefighters to protect themselves and be prepared to use all available personal protection equipment and breathing apparatus.

Flame, Heat, and Smoke

Direct contact with flames can result in devastating skin burns and serious damage to respiratory tract. To prevent skin burns during a fire, personnel should maintain a safe distance from the fire unless they are wearing proper personal protective gear and equipped with firefighting equipment to combat the fire. The firefighter's ensemble and breathing apparatus are worn by the firefighting team. The firefighter's ensemble will protect against flames and heat while the breathing apparatus will protect the respiratory tract.

A single fire will produce temperatures in excess of 2000°F (1093°C) specifically when a ship's compartment has flashed over. Temperatures above 150°F (66°C) are hazardous to humans and surface temperature as low as 160°F (71°C) will result in second degree burns if contact is maintained for 60 seconds. The dangerous effects of heat produce from fires can range from minor injury to death. Direct exposure to heated air may cause an increased heart rate, dehydration, heat exhaustion, burns and blockage of the respiratory track by fluids. So it is very imperative that the firefighting team wear protective gear. Even in protective gear the firefighter is at risk. A firefighter exposed to excessive heat over an extended period of time could develop hyperthermia, a dangerously high fever that can damage nerve centers. If heat management techniques are not use during fighting the fire, a firefighter may be seriously injured or death may even occur.

Smoke is a visible product of fire that increases problems for breathing and hinders sight. Smoke consists of water vapors, acids, carbon, various chemicals and other unburned substances in the form of suspended particle which can be poisonous or irritating when inhaled. Smoke greatly reduces visibility in the vicinity of the fire and it irritates the eyes, nose, throat and lungs. The presence of smoke or irritation to the nose and eyes indicates the presence of hazardous fire gases. Breathing a low concentration for an extended period of time or a heavy concentration for a short time will cause great discomfort. The firefighting team must activate breathing apparatus where smoke is present, when irritation to nose and eyes indicates the presence of hazardous fire gases, and in the affected fire area to prevent exposure to hazardous smoke and toxic gases. To cope with reduced visibility from the smoke, the firefighter can wear a headlamp on their helmet and use the thermal imager to fine the seat of the fire.

Gases

The specific gases produced by fires will depend on the substance being burned. Some of the gases produced by fires are toxic while other gases are nontoxic but are still dangerous in additional ways. The common gases produce is carbon monoxide (CO) which is the product of incomplete combustion and CO_2 which is the product of complete combustion. In a smoldering fire, the ratio of carbon monoxide to carbon dioxide is typically greater than it would be in a free-burning fire that is well-ventilated. Other common gases that are produced by shipboard fires are hydrocarbon vapors such as hydrogen chloride (HCl), hydrogen cyanide (HCN), and hydrogen sulfide (H_2S).

CO is produced in fires when there is not enough oxygen present for the complete combustion of all of the carbon in the burning material. CO is a colorless, odorless, tasteless, and nonirritating gas that can cause death even in small concentrations. Personnel who are exposed to CO concentrations of 1.28 percent will become unconscious after two or three breaths and possibly die in 1 to 3 minutes if left in the affected area. CO also has a wide explosive range. An open flame or spark will set off a violent explosion if CO is mixed with the atmosphere in the amount of 12.5 to 74 percent by volume.

A fire will produce CO_2 when there is complete combustion of all of the carbon in the burning material. CO_2 is a non-toxic colorless and odorless gas that will cause unconsciousness after prolonged exposure at 10 percent volume and higher. Unconsciousness can occur in 1 minute or less when CO_2 levels are above 11 percent volume and death could occur in a sufficient quantity. The danger of asphyxiation should never be taken lightly because CO_2 is an inert gas that displaces oxygen and does not give any warning of its presence especially in dangerous amounts. Because CO_2 does not support combustion and it does not form explosive mixtures with any substances, these characteristics are what make CO_2 very effective as a fire-extinguishing agent.

H_2S is produced during fires where halon is used as the extinguishing agent. H_2S is a decomposed product of halon and it is also produced by the rotting of foods, cloth, leather, sewage, and other organic materials. Firefighters must use caution when fighting fires around sewage systems and in spaces where there has been a sewage spill. H_2S is a colorless gas that smells like rotten eggs. A flame or spark can cause a violent explosion when the atmosphere contains 4.3 to 46 percent. Concentrations of H_2S as low as 20 parts per million (ppm) is extremely poisonous by causing personnel to quickly become unconscious, stop breathing, and possibly die after one breath in an atmosphere that contains 1,000 to 2,000 ppm of H_2S.

Insufficient Oxygen

A fire in an enclosed shipboard compartment may cause the supply of oxygen to diminish due to an enormous amount of oxygen is used by the fire leaving barely no oxygen to breathe. Approximately 20.8 percent is the breathable amount of oxygen normally present in the atmosphere. Personnel will breathe and work best with this amount of oxygen during normal routine. Human muscular control will be reduced once the oxygen content drops to about 16 percent. Personnel will start feeling fatigue and their judgment become impaired at about 10 percent to 14 percent oxygen. When oxygen concentrations fall below 10 percent personnel will become unconsciousness. During periods of intense firefighting efforts, the firefighter's body requires more oxygen and increased demands may result in oxygen deficiency symptoms at normal oxygen levels. This is why it is important for the firefighting team to wear a breathing apparatus and rotate positions.

Fire Extinguishment

Fires may be extinguished by removing one side of the fire triangle (fuel, heat, or oxygen), slowing down combustion, or interrupting the uninhibited chemical reaction. The firefighting team must remove the fuel, heat or oxygen by using any specific methods which solely depends on the fire classification and the circumstances surrounding the fire.

Removing Fuel

Although it is not usually possible to remove the fuel to extinguish a fire, there may be circumstances in which it is possible to remove the fuel. If burning material can safely be heaved over the side of the ship, then the firefighter will do it as soon as possible. This process is normally done on flight decks and weather decks. Firefighting teams must be ready at all times to shift combustible materials to safe areas and take whatever possible measures to keep additional fuel hazards away from the fire such as immediately isolating all supply valves in fuel oil, lube oil, and JP-5 piping systems.

Removing Heat

The fire will be extinguished if there is enough heat removed from cooling the fuel below its ignition temperature. Heat from a fire is transferred in the following ways:

1. Radiation.
2. Conduction.
3. Convection.

Heat is emitted through the air in all directions during the radiation process. This process is what causes a person to feel hot when they stand near an open fire because the heat travels outward from the fire in the same manner as light. Heat is absorbed, reflected or transmitted whenever it contacts an object. Absorbed heat will increase the temperature of the absorbing object. A clear example of this process would be radiant heat that is absorbed by the ceiling will increase the temperature of the ceiling, perhaps enough to ignite its paint. Heat will radiate in all directions unless it is blocked. Radiant heat will extend a fire by heating all combustible objects in its path, causing the objects to produce vapors, and then igniting the vapors. Shipboard fires can spread as a result of radiating bulkheads and decks. Intense radiated heat can make an approach to the fire extremely difficult. For this reason, protective clothing should be worn by firefighters.

During the process of conduction, heat is transferred through an object or from one object to another by direct contact from molecule to molecule. A simple example is a pot of water on a hot stove. Heat is conducted through the pot to water. This is way the water begins to boil. So a ship's thick steel wall or bulkhead with a fire on one side can conduct heat from the fire and transfer the heat to the adjoining compartments. Heat transfer by conduction is hazardous since the fire will move from one deck to another and one compartment to another. This is why immediately setting fire boundaries in the adjoining compartments is extremely important.

Convection is the transfer of heat through the motion of circulating gases or liquids. During this process, the air and gases rising from a fire are heated. Heat is transferred by convection through the motion of smoke, hot air and heated gases produced by a fire. These gases transfer the heat to other combustible objects that are within reach. An example of convection is taking that same pot of water on a stove. As the water at the bottom of the pot is heated up, the heated water moves to the top bringing the cooler water to the bottom of the pot causes a continuous circulation. This is why rolling bubbles appear in the pot of water. Heat transferred by convection is particular dangerous in the ship's ventilation systems. These ventilation ducts may carry heated gases from the fire to other locations several compartments or decks away from the fire area. If there are combustibles with a low flash point within a compartment served by the same ventilation system, a new fire will ignite.

Since heat is transferred in several methods the firefighter must eliminate the heat side of the fire triangle, cool the fire by applying firefighting agents that will absorb the heat. Although several agents serve this purpose, water is the most commonly used cooling agent and ships have an inexhaustible supply of it through seawater pumps. Firefighting water may be applied in the form of a solid stream, as a fog, or used together with AFFF.

Controlling Oxygen

Oxygen is the one of the components of the fire triangle. Oxygen is difficult to control because there is no way possible to remove the oxygen from the atmosphere that normally surrounds a fire, but oxygen can be diluted or displaced by other substances that are noncombustible. For example, in an open area such as the weather decks or flight deck if CO_2 is used the agent would be blown away from an open deck area, especially if the ship is out to sea or during windy conditions. However, a fire in a trash container can be extinguished by placing a lid tightly over the container, blocking the flow of air to the fire, or it could be extinguished by CO_2 dilution. When the lid is used the fire consumes the oxygen in the trash container, it becomes starved for oxygen and the fire is extinguished. If CO_2 is used, then the fire in the trash can's oxygen is displaced with the CO_2 and the fire is extinguished.

A fire in a ship's compartment is abandoned by closing the door and other openings to reduce the supply of fresh oxygen. This will reduce the burning rate of the fire, thus preventing flashover and may even reduce the fire to a smoldering condition. In an enclosed compartment such as an engine module can be extinguished by diluting the air with CO_2 gas. This dilution must proceed to a certain point before the flames are extinguished. The point at which the dilution is enough to extinguish the fire can be reached faster if all ventilation systems is secured to the space. In general, a large enough volume of CO_2 must be used to reduce the oxygen content to 15 percent or less and it is important for the firefighter to know when to use CO_2 as the preferred extinguishing agent.

Reducing the Rate of Combustion

Flaming combustion occurs in a sophisticated series of chemical chain reactions. Once the chain reaction sequence is disrupted, a fire will be extinguished immediately. The firefighting agents commonly used to attack the chain reaction and inhibit combustion are dry chemicals, halons, and heptafluoropropane (HFP). These chemical agents directly attack the molecular structure of compounds formed during the chain reaction sequence and breaking down these compounds will adversely affect the flame-producing capability of the fire. The attack is extremely rapid and these chemicals work by interrupting the chemical chain reaction which reduces the rate of combustion and the fire is extinguished quickly. The firefighter must take in consideration that these agents will not cool a smoldering fire or combustible liquids whose container has been heated above the liquids' ignition temperatures. In these situations, the extinguishing agent must be maintained on the fire until the fuel has cooled. The firefighting team must use a cooling medium such as water or AFFF to cool the smoldering embers.

Extinguishing Agents

There are many substances that may be used as extinguishing agents. The agent or agents that are used in firefighting evolutions will solely depend upon the classification of the fire and the circumstance. The agents can be utilized in fire protection systems or portable fire extinguishers. The firefighting agents that are the commonly used aboard naval ships are the following:

1 Water.
2 Aqueous Film Forming Foam (AFFF).
3 Carbon Dioxide (CO_2).
4 Halon.
5 Potassium Bicarbonate ($KHCO_3$).
6 Aqueous Potassium Carbonate (APC).
7 Heptafluoropropane (HFP).

Water

Water is a cooling agent onboard U.S. Navy ships and primarily used to extinguish Class A fires. The ships use an inexhaustible supply of water from the sea through the seawater pumps called fire pumps that supply the firemain system. Water extinguishes a fire by lowering its surface temperature below the ignition temperature of the fuel. Water is mostly effective when it absorbs enough heat to raise the water temperature to 212°F (100°C) which causes the seawater to absorb more heat as it changes to steam. The steam carries away the heat and results in the lowering of the temperature of the surface. Navy ships use both fresh water and seawater in firefighting systems. Seawater is used in various sprinkler systems such as berthing sprinkler systems, magazine sprinkler systems and fireplugs that provide water to firefighting hoses throughout the ship. Fresh water is used on some ships for small diameter quick response hose reels and the water mist system.

Water in the form of straight stream or solid stream is used to reach into smoke-filled spaces or areas at a distance to help protect the firefighter. When a straight stream is needed as an extinguishing agent, the stream should be directed into the seat of the fire. For maximum cooling, the water must come in direct contact with the burning material. A straight stream is most commonly used to break up and penetrate materials. Firefighters should use extreme caution when using straight stream water because it is a conductor of electricity. Firefighters should not use it on electrical fires and reframe from direct contact with electrical components.

Water in the form of water fog is very effective for firefighting purposes. The water fog can be used in low or high velocity depending on the circumstance. Low velocity fog pattern is utilized to provide protection to firefighters from convective and radiant heat. High velocity fog pattern is utilized for firefighting and it is more effective than the straight stream pattern. When using high velocity fog pattern, the fog must be applied directly to the area to be cooled if its benefits are to be fulfilled.

Aqueous Film-Forming Foam (AFFF)

Foam is a highly effective extinguishing agent for smothering large fires, specifically oil, gasoline, and jet fuel fires. AFFF, sometimes called "light water," is synthetic, film-forming foam designed for use in shipboard firefighting systems. The foam proportioning/injection equipment generates a white foam blanket that is unique because it floats on top of the flammable surface. AFFF concentrate is a clear to slightly amber colored liquid. The AFFF solution of water and concentrate possesses a low viscosity and is capable of quickly spreading over the flammable liquid surface. AFFF concentrate is nontoxic and biodegradable in diluted form.

AFFF is mainly used to fight Class B (flammable liquids) but can be used on Class A fires as well. It is very effective on Class A fires because it contains seawater. Navy's military specifications authorize a 6 percent AFFF concentrate and 3 percent AFFF concentrate. The ratio of 6 percent (6 parts) is mixed with 94 percent seawater (94 parts water), whereas the ratio of 3 percent (3 parts) is mixed with 97 percent seawater. When AFFF is properly proportioned with water, it provides three fire extinguishing advantages. The first advantage is AFFF will form an aqueous film on the surface of the fuel which prevents the escape of the fuel vapors preventing reflash of the fire. The second advantage is AFFF forms a layer of foam over the fuel which effectively excludes oxygen from the fuel surface. The third advantage is AFFF is mixed with water to provide a cooling effect.

Carbon Dioxide (CO_2)

CO_2 is a dry, noncorrosive gas that is heavier than air and remains close to the surface. CO_2 will become inert with most substances and it does not leave residue nor damage machinery or other equipment. CO_2 is a nonconductor of electricity that can safely be used to fight fires that might present electric shock hazards. CO_2 is used in portable fire extinguishers, fix flooding systems, hose reel systems, and transfer units in shipboard applications. Although CO_2 is nonpoisonous, it is dangerous because it causes asphyxiation and should not be used to extinguish a fire in a small or confined compartment without the use of a breathing apparatus.

CO_2 extinguishes the fire by diluting and displacing the oxygen supply and when it is directed into a fire the sufficient oxygen to support combustion is removed causing the flames to diminish. A fire will use 21 percent oxygen to continuously burn and when CO_2 is utilized the oxygen supply will be reduced below 15 percent. Some ordinary combustible Class A fires requires the oxygen content be reduced to less than 6 percent in order to extinguish glowing combustion (smoldering fire). CO_2 has limited cooling capabilities that may not cool the fuel below its ignition temperature and it is more likely to allow reflash than other extinguishing agents. In this case, firefighters must remember to standby with additional backup extinguishers or use an additional extinguishing agent such as water.

Aircraft carriers and other large ships have a CO_2 transfer unit to refill portable CO_2 extinguishers. This system transfers CO_2 from a 50-pound cylinder into the portable 15-pound extinguisher cylinder. The system is installed in a well-ventilated area. The CO_2 transfer unit consists of an electric motor, a pump, a high-pressure hose, a control valve, adapters, and fittings. The transfer unit is maintained according to the planned maintenance system and with the manufacturer's instructions.

Halon

Halon is a halogenated hydrocarbon, which means that one or more of the hydrogen atoms in each hydrocarbon molecule have been replaced by one or more atoms from the halogen series such as fluorine, chlorine, bromine, or iodine. This substitution provides non-flammability and fire extinguishing properties that is an effective agent against Class A, Class B, and Class C fires. Halon extinguishes fires by interrupting the chemical chain reaction of the fire. The two types of halon used aboard naval ships are halon 1301 and halon 1211.

Halon 1301 is known chemically as bromotrifluoromethane which consists of one atom of carbon, three atoms of fluorine, no chlorine atoms, one bromine atom, and no iodine atoms. It is a colorless, odorless gas with a density approximately five times that of air that does not conduct electricity or leaves a residue. It is stored in a cylinder as a compressed liquid that is super-pressurized with nitrogen. The cylinders are used in fixed flooding systems for extinguishing flammable/combustible liquid fires in engine rooms, fuel pump rooms, and hazardous material storage rooms.

Halon 1211 is chemically known as bromochlorodifluoromethane which consists of one atom of carbon, two atoms of fluorine, one atom of chlorine, and one atom of bromine. It is also colorless like halon 1301 but has a sweet smell. This agent is stored and shipped as a liquid and pressurized with nitrogen gas which is necessary since the vapor pressure is too low to convey it properly to the fire area. This agent is not used in fixed flooding systems because of its lower volatility and high liquid density which enables the agent to be sprayed as a liquid and therefore propelled into the fire zone to a greater extent than is possible with other gaseous agents. Halon 1211 is used in 20-pound capacity portable fire extinguishers on air cushion landing craft (LCAC) and flight decks of large air-capable ships.

Halon is a very effective agent but for it to function properly as an extinguishing agent, it must decompose. It decomposes upon contact with flames that are approximately 900°F (482 °C). Once the agents decompose several toxic gases such as hydrogen fluoride (HF) and hydrogen bromide (HBr) are produced. Both gases are irritating to the eyes, skin, and upper respiratory tract as well as chemical burns. Personnel should immediately evacuate from a compartment where halon has been released unless they are wearing a breathing apparatus and protective clothing. It has been discovered that halon chemicals have a substance that depletes the ozone layer. Newer naval ships have been installing alternative firefighting systems where halon application is needed. These other alternatives will be discussed later in this chapter and the next chapter.

Potassium Bicarbonate ($KHCO_3$)

Dry chemical powders extinguish a fire by a unique chemical mechanism as this agent does not smother the fire nor does it provide a cooling effect. Dry chemical powders disrupt the chemical chain reaction of the fire by suspending the molecules in the fire by placing a temporary separation between the heat, oxygen, and fuel. This process maintains this separation affect just long enough for the fire to be extinguished. Several types of dry chemicals are used as fire extinguishing agents, but Navy ships use one of the most important dry chemical powders called potassium bicarbonate ($KHCO_3$) which is known as Purple-K-Powder (PKP). PKP is used to extinguish Class B and Class C fires because it is very effective against these fires. Because PKP is corrosive and abrasive, it should only be used on Class C fires during emergencies when other firefighting agents have been exhausted or not available. PKP is available in 18-pound and 27-pound portable extinguishers on naval ships. PKP can be used in conjunction with AFFF to extinguish large Class B fires.

Aqueous Potassium Carbonate (APC)

Grease fires are classified as a Class B fires on U.S. Navy ships, whereas civilian firefighting classifies it as class K. APC (K_2CO_3) is used on board Navy ships in galleys for extinguishing burning cooking oil and grease in deep fryers and its ventilation exhaust ducts. APC solution consists of 42.2 percent potassium carbonate and 57.8 percent water. When APC is used in combating Class B fires involving cooking grease and oils, the application of alkaline solution generates a soap-like froth that will remove oxygen from the surface of the grease or oil. By removing the oxygen leg of the fire triangle, the grease fire is extinguished.

Heptafluoropropane (HFP)

HFP is the Navy's term for a specific gaseous fire extinguishing agent which is an alternative to halon 1301 in some of the newer ships. HFP consists of several compounds such as carbon, fluorine, and hydrogen in the formula C_3F_7H. It is also known commercially as HFC-227ea and by the proprietary trade names FE-227 (E.I. Du Pont De Nemours & Co) and Firemaster 200 (or FM-200, Great Lakes Chemical Co.). HFP is a colorless, odorless and electrically non-conducting gas that is clean and leaves no residue. HFP is stored in steel containers at 600 PSIG at 70°F (41 bars at 21°C) as a liquefied compressed gas that uses nitrogen to improve the discharge characteristics. When HFP is discharged the liquid will vaporize into a gas at the discharge nozzle and it is evenly distributed as it enters the affected shipboard compartment.

HFP is used in fire extinguishing systems for Class B flammable liquid and combustible liquid pool fires and spray fires. It is installed where water mist system applications are unsuitable and considered less effective such as flammable liquid storerooms that stores flammable liquids with low-flash points. HFP has replaced halon 1301 in nuclear-powered aircraft carrier (CVN) 76 and 77 class ships' fuel pump rooms and flammable liquid storerooms as well as landing helicopter deck (LHD) 8 and landing platform dock (LPD) 17 class ships' engine enclosures, the hazardous materials complex, and flammable liquid storerooms.

HFP extinguishes the fire by replacing the oxygen of the fire. HFP decomposes upon contact with flames or very hot surfaces above 1300°F (700°C) and produces hydrogen fluoride (HF) acid gas in very high concentrations. HFP's design concentration and short discharge time (ten seconds maximum) are intended to provide rapid extinguishment and minimize the formation of HF; however, the atmosphere in the compartment after discharge should be considered extremely toxic for the firefighting team even equipped with fire protective clothing and breathing apparatus. Procedures have been developed for firefighting teams to follow when entering compartments that are protective by HFP and potential high concentrations of HF after discharge.

Portable Fire Extinguishers

Fires can occur at a moment notice and early response can prevent a fire from spreading throughout the ship. Portable fire extinguishers are used as an effective early response to a developing fire (Figure 3-3). They are strategically placed throughout the Navy ships to aid in rapid response.

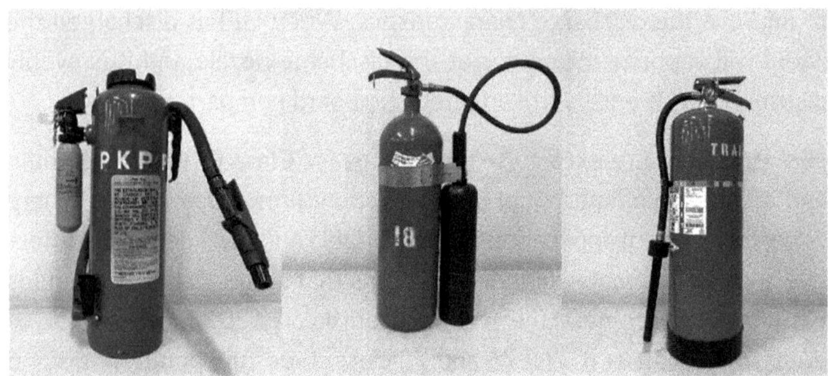

Figure 3-3. Portable Fire Extinguishers (PKP, CO_2, AFFF)

The DCPO mention in chapter 2 is responsible for ensuring that they are ready to be used by conducting periodic inspections and maintenance on all portable fire extinguishers. The portable fire extinguishers that are used aboard all Navy ships are as follows:

1. Dry chemical (PKP).
2. Carbon dioxide (CO2).
3. Aqueous film-forming foam (AFFF).

Dry Chemical Extinguisher

Portable dry chemical extinguishers are used primarily on Class B fires. Potassium bicarbonate ($KHCO_3$) or Purple-K-Powder (PKP) is the chemical used in these extinguishers on Navy ships. The dry chemical dispensed from the extinguisher stops flaming combustion by disrupting the chemical chain reaction of a fire. As previously mentioned, PKP is an effective agent used on Class C fires, but it causes damage in electrical and electronic components; therefore, carbon dioxide is preferred agent. Additionally, PKP will damage gas turbines or jet engines, but if it is absolutely necessary it may be used to extinguish internal fires of gas turbines or jet engines. PKP is not as effective on Class A and large Class B fires and should only be utilized to knock down flames and keep the fire under control until an appropriate extinguishing agent is used such as seawater on Class A fires and AFFF on Class B fires.

PKP extinguishers are found on naval ships in 18-pound and 27-pound size bottles. The 18-pound extinguisher has an effective range of 19 feet with a continuous discharge time of 10 seconds and they are located near flammable storage lockers, paint lockers, and galleys. The 27-pound extinguisher has an effective range of 21 feet with a continuous discharge time of 11 seconds and they are located in engine rooms, auxiliary rooms, fuel pump rooms, hangar bays, and flight deck catwalks. PKP extinguishers have a CO_2 cartridge that is mounted on the outside of the extinguisher shell except for the 27-pound LEHAVOT extinguisher which has the CO_2 gas pressure cartridge mounted inside the shell. This CO_2 cartridge provides the propellant charge for the extinguisher.

The PKP extinguisher should not be pressurized until it is ready to use. When it is ready to use ensure to adhere to the following steps:

1 Remove the extinguisher from the mounting bracket
2 Remove the seal and pull the locking pin from the puncture lever marked PUSH.
3 Push the puncture lever down to penetrate the seal of the CO_2 cartridge. The extinguisher is now ready to use.
4 Test agent by quickly squeezing and releasing the grip on the nozzle.
5 Carry the extinguishers to the scene of the fire, if outside ensure to approach it from the windward side, if possible. Hold the extinguisher in one hand and the nozzle in the other hand.

6 Discharge the PKP by squeezing the squeeze grip on the nozzle. Hold the nozzle firmly and direct the extinguishing agent at the base of the fire. Use a wide-sweeping motion from side to side. This will apply a dense, wide cloud of dry chemical over the area. Remember that the 27-pound extinguisher has a 21-foot range and the 18-pound extinguisher has a reach of 19 feet.

7 Be certain that all of the fire in the area is extinguished before moving in farther. If the fire appears to be too large or if there is a possibility of being outflanked or surrounded by flames, attack the fire with the assistance of two or more personnel using extinguishers.

8 Do not try to economize on the dry chemical instead use as much as necessary and as many extinguishers as necessary to extinguish the fire completely.

9 Always use seawater or AFFF in conjunction with the extinguisher or as a backup.

10 After the fire has been extinguished, the PKP extinguisher must be inverted and the discharge lever of the nozzle must be squeezed to release residual pressure, and the nozzle must be tap on the deck to release residual powder. Then turn into the responsible DCPO to replace the CO_2 cartridge, refill it with PKP, and placed the extinguisher back at its location for future use.

AFFF Extinguisher

Portable AFFF fire extinguishers are used to provide a vapor seal over a small fuel spill, extinguish small Class B fires such as fire in deep fryers, and for standing fire watch during welding and cutting jobs. This extinguisher is located throughout the ship in DCRS, galleys, and near paint lockers and hazardous material storage lockers. The portable AFFF fire extinguisher is a stainless steel cylinder containing 2 1/2 gallons of premixed AFFF concentrate and water. It is pressurized with compressed air to 100 psi at 70°F and weighs approximately 28 pounds when fully charged. When AFFF is dispensed the mixture will expand about 6.5 to

1 and will produce about 16 gallons of foam. The AFFF extinguisher has a 55-65 second continuous discharge time and discharge stream range of 15 feet which decreases during discharge.

The AFFF extinguisher is designed for use on Class B pool fires and it may also be used on Class A fires. AFFF is not recommended for use on Class C fires because energized electrical components can be damaged and cause electrical shock to the user. AFFF extinguishes Class A fires by cooling the substance on fire. It is superior to water because AFFF has added wetting and penetrating ability. AFFF extinguishes a Class B fire or protects an unignited fuel spill by floating on the top of the flammable liquid and forming a vapor seal that traps the fumes. One AFFF extinguisher will effectively extinguish 20 square feet (4 1/2 feet by 4 1/2 feet) of flammable liquid fire. When operating the AFFF extinguisher the user must start approximately 15 feet away from the fire using a sweeping motion and directing the AFFF at the base of the fire. To prevent a fire, one AFFF extinguisher can provide a vapor seal on a fuel spill up to 40 square feet (about 6 feet by 6 feet) in size. Larger fuel spills, spills which are not fully accessible or visible, should be covered with foam using 1-1/2-inch AFFF hose or by the installed AFFF bilge sprinkling system which will be discussed in the next chapter.

Deep fryer fires often require special procedures to extinguish them such as using combinations of AFFF and PKP to extinguish these fires and prevent it from spreading throughout the compartment or into ventilation ducting. The user must take safety precautions by not directing the AFFF stream straight into the hot cooking oil because doing so can result in immediate boiling of the AFFF. This violent boiling may result in hot cooking oil splashing out of the fryer onto the user or firefighters. The AFFF stream should only be directed at the back wall of the fryer, allowing the stream to flow onto the surface of the burning oil. This technique does not cause the cooking oil to splatter allowing the fire to be extinguished and a layer of foam formed over the oil.

The AFFF extinguisher is immediately ready to use. When it is needed the user shall conduct the following steps:

1 Remove the extinguisher from the mounting bracket.
2 Ensure that extinguisher's dial or needle is pointing in the green.
3 Remove the seal and pull the locking pin from lever.
4 Test agent by quickly squeezing and releasing the grip on the lever.
5 Carry the extinguishers to the scene of the fire, if outside ensure to approach it from the windward side, if possible. Hold the extinguisher in one hand and the nozzle in the other hand.
6 Discharge the AFFF by squeezing the squeeze grip on the lever. Hold the nozzle firmly and direct the extinguishing agent at the base of the fire. Use a wide-sweeping motion from side to side. If used during a deep fryer fire ensuring to aim the flow to the back wall of the deep fryer.
7 Be certain that all of the fire in the area is extinguished before moving in farther. If the fire appears to be too large or if there is a possibility of being outflanked or surrounded by flames, attack the fire with the assistance of two or more personnel using extinguishers.
8 Do not try to economize on the agent instead use as much as necessary and as many extinguishers as necessary to extinguish the fire completely.
9 After the fire has been extinguished, the extinguisher is turned over to the responsible DCPO to be refilled, recharged and placed back in its location for future use.

Carbon Dioxide (CO_2) Fire Extinguisher

The standard Navy CO_2 fire extinguisher has a rated capacity of 15 pounds of CO_2 (by weight) and it extinguishes a fire by removing the oxygen. The maximum range of a 15-pound CO_2 extinguisher is 4 to 6 feet from the outer end of the horn with a continuous operation of 40 to 45 seconds. CO_2 extinguishers are primarily used on small electrical fires (Class C), small Class A fires and have limited effectiveness on Class B fires. The CO_2 extinguishers are normally located within 30 feet of equipment with a high potential for electrical fire. Compartments that are normally outfitted with portable CO_2 extinguishers are shops, DCRS, berthings, flight deck catwalks, compartments containing electrical motors, switchboards and panels, electronic and navigational areas, weapons cleaning and fire control areas, galleys, and machinery spaces.

The user of the CO_2 extinguisher must be knowledgeable of the precautions when utilizing the extinguisher. CO_2 is an inert gas that displaces oxygen and can cause asphyxiation to the user if utilized in small enclosed or confined spaces in which the user must ensure to use a breathing apparatus. The discharge of CO_2 causes static electricity and poses an electrical shock to the user if they do not ground the bottle by placing it on the ship's deck or frame. Also the static electricity can cause an explosion or fire if it is used in a flammable vapor enriched environonment or near flammable substances.

The CO_2 extinguisher is immediately ready to use. When it is needed the user shall conduct the following steps:

1. Remove extinguisher from mounting bracket.
2. Place the CO_2 extinguisher on the deck and remove the locking pin from the valve.
3. Test extinguisher by pointing the nozzle to the deck and rapidly squeeze and release the lever ensuring that the agent has a positive discharge.
4. Place the locking pin in lever and carry the extinguisher in an upright position and to the scene of the fire getting as close to as possible.
5. Place the extinguisher on the deck and remove the locking pin from the valve.
6. Grasp the insulated handle of the horn. Rapidly expanding CO_2 causes the horn to become quite cold.
7. Squeeze the operating lever to open the valve and release CO_2 directly towards the base of the fire.
8. Do not try to economize on the agent instead use as much as necessary and as many extinguishers as necessary to extinguish the fire completely.

9 After the fire has been extinguished, the extinguisher is turned over to the responsible DCPO to deliver it to the shop responsible to get the bottle hydrostatic tested and refilled. Once it gets refilled the responsible DCPO must place it back in the proper location for future use.

Self-Contained Breathing Apparatus (SCBA)

The SCBA is designed for firefighting evolutions. The SCBA is designed to be entirely self-contained allowing the user to breathe independently of the outside atmosphere. Clean breathable air is provided from a harness and cylinder carried on the back of the user which allows the user to enter oxygen deficient spaces or toxic atmospheres caused by fires and during chemical spills. The SCBA operates using a pressure-on-demand system that provides air to the wearer when the wearer inhales. When the wearer inhales, the regulator will open allowing air to flow into the mask. When the wearer exhales, a simple one-way valve known as an exhalation valve allows the exhaled air to escape without allowing contaminants from the outside atmosphere to enter the mask. This operation is also known as an open-circuit system.

The 30-minute cylinder and 45-minute cylinder are two types of cylinders used onboard ships. The 30-minute cylinders are used for such investigators, electricians, boundarymen, and smoke controlmen while the 45-minute cylinders are used for members of the firefighting team. Each cylinder is charged to 4,000 to 4,500 psi with grade D oxygen and both will provide the wearer with either 30 minutes or 45 minutes of oxygen but it could be longer or shorter depending on the demand of the wearer. All SCBAs have four basic component assemblies and operate in a similar manner. The assemblies are as follows:

1 Harness/backpack assembly.
2 Air tank assembly.
3 Regulator assembly (including low-pressure alarm and high-pressure hose).

4 Face piece assembly with heads up display and vibration alert system.

The SCBA's cylinders can be refilled utilizing SCBA refill stations throughout the ship that provides grade D oxygen. Also, spare bottles can be used to replace emptied cylinders by the utilizing a designated changed out area authorized by the DCA and shipboard procedures. The SCBA is an important piece of protective equipment and personnel must be given proper training on its safe operation. This training will be conducted according to the manufacturer's technical manuals, and maintenance will be accomplished according to the associated planned maintenance system procedures and the manufacturer's technical manual. The training will include but not be limited to the following:

1 SCBA usage.
2 SCBA cleaning.
3 SCBA inspection.
4 SCBA replacement of components.

Emergency Escape Breathing Device (EEBD)

Smoke and various toxic gases will be produced during a fire and personnel who initially reported the fire must be aware of this. The Rapid Response team must be aware of any hazardous materials stored within the compartment. All shipboard personnel must know when to evacuate the affected compartment and allow the firefighting and emergency response team to handle the casualty. Personnel must be trained to utilize devices to aid their evacuation. Throughout the ship, devices known as EEBDs are stored in marked locations in most compartments. These are used to provide a limited amount of breathable oxygen to aid personnel with evacuating the affected area to a safe zone. The EEBD shall not be utilized to combat a fire or investigate a toxic hazard. When the ship is out to sea, machinery and auxiliary spaces are operated by personnel known as watch standers and they are trained to take initials actions to combat fires. During their initial firefighting actions, they are trained to grab an EEBD and carry it over the shoulder by the strap attached to the carrying case. Carrying the EEBD over the shoulder will allow the watch standers to immediately escape once they are overcome with smoke or lose control of the fire.

The EEBD provides the user with 10 minutes of oxygen for escape, even though it is rated up to 32 minutes. It has a service life of 15 years. It is a one-time use device and it is properly disposed after each use. It uses compressed oxygen in automatic on-demand regulated system. As previously mention, the strap on the case allows the EEBD to be worn on the shoulder as well as worn on the belt. The EEBD cases are mounted in stationary brackets in various locations throughout the machinery and auxiliary spaces. Berthing or sleeping quarters have the EEBD cases mounted inside the rack or bed area. Offices and workshops have them mounted in their spaces as well.

The EEBD can be put on in a few seconds by simply unlatching the case, pulling out the unit, inserting the mouthpiece and attaching the nose clip to the nose. The attached hood can be used at any time as needed during emergency egress. The hood protects the user from a smoke filled and hazardous environment while allowing a full range of view. The Teflon hood and breathing bag provide excellent heat and chemical resistance. The compressed oxygen and mouthpiece combination allows the EEBD to be put on in a smoke-filled environment. The actual EEBD can be identified by its orange storage case. The training EEBD is equipped with two extra mouthpieces and it is stored in a light blue storage case. The training EEBD provides training in both the worn and stored positions and to avoid confusion during an actual emergency, it should be locked or stored away.

Firefighter's Ensemble

The firefighter's ensemble (FFE) is designed to protect the firefighter from heat, falling debris, and short durations of flame exposure (Figure

3-4). Sailors' safety during actual casualties and training evolutions may depend on proper wearing of the ensemble. To ensure rapid deployment of the firefighting team, the team will practice quickly dressing up in the ensemble with the SCBA.

Figure 3-4. Firefighter's Protective Ensemble (FPE) with SCBA

The firefighter's ensemble with SCBA consists of the following items:

1 Firefighter's coveralls.
2 Firefighter's anti-flash hood.
3 Firefighter's helmet.
4 Firefighter's gloves.
5 Firefighter's boots.
6 Anti-flash clothing.

The FFE is not a proximity suit; therefore, it is not designed as protective gear during rescue operations during a fire caused by an aircraft crash. Prolonged contact with flames may cause the clothing to transmit dangerous heat to the body or may cause the clothing to burn which could result in serious injury or death to the firefighter. The ensemble does not offer complete protection against CBRN effects nor fragments from exploding ordnances.

Firefighter's Coveralls

The firefighter's coverall is a one-piece jump suit style that consists of an outer shell, a vapor barrier, and an inner fire-retardant liner. The knees, bottoms of the thigh pockets, and bottoms of the legs are reinforced with leather for extra protection of the legs. It is outfitted with reflective marking strips around the upper arms, lower legs, and torso to highlight the outline of the fire fighter so they can be visible in dense smoke or dim light.

Firefighter's Anti-Flash Hood

The firefighter's anti-flash hood provides protection to the head, neck, and face except the eyes, but wearing the SCBA face piece will protect the eyes. The hood can be worn with the SCBA over the straps of the face piece. It has a face opening which is elastic to fit snuggly around the SCBA face piece. The hood is available in a single size which fits all.

Firefighter's Helmet

The firefighter's helmet is designed to protect the head, neck, and face from, heat, falling objects, and short duration of flame exposure. The helmet shell material is made of heat-resistant fiber glass and equipped with a rear brim, face shield, chin strap, adjustable suspension, reflective markings, and ear flaps that cover the side of the head and neck.

The helmet's used for firefighting have high-intensity battery-powered lights that are attached to allow the firefighter to see through smoke, darkness, and dim lighted areas.

Firefighter's Gloves

Firefighter's gloves are designed to protect the wearer against abrasion, heat, and short duration of flame exposure. They are made of leather, aluminized fabric with a waterproof vapor barrier and a fire-retardant liner. The gloves are available in various sizes to properly fit and aid the wearer ability to operate the nozzle and maneuver the firefighting hose.

Firefighter's Boots

The firefighter's boots are fabricated from black rubber with insulated soles and non-skid aggressive treads on the bottom to prevent firefighters from slipping on wet or oily decks. They are equipped with steel toes, puncture-proof insoles, and some models have steel chin guards to protect the firefighter's feet and legs. They are available in a variety of sizes and two models such as the knee-high and hip-length. The knee-high version is the most common boot used for firefighting. The hip boots will be used in situations that may require the firefighter to enter deep water that is hot or boiling.

Anti-flash Clothing

Anti-flash clothing is used to protect personnel from transient high temperatures that may occur from the use of high explosive weapons and from being burned in a fire. Anti-flash clothing items are known as coveralls and socks made from fire-retardant material. The coveralls and socks are available in various sizes. The fire-retardant coveralls are to be worn underneath the firefighter's coverall and the socks are to be worn in the firefighter's boots.

Navy Firefighters Thermal Imager (NFTI)

The thermal imager is a device like a camcorder that allows the user to see through dense smoke and light steam by sensing the difference in infrared radiation given off by objects with a temperature difference of at least 0.05 to 5 degrees depending on the NFTI's model or modification. It can be used to for the following:

1 Investigate reported fires.
2 Locate the seat of the fire.
3 Locate and guide rescuers to injured personnel.
4 Set and maintain fire boundaries.
5 Locate ignition sources during fire overhaul.

The NFTI has gone through several models and modifications. The current model used is the Talisman. The NFTI utilizes self-contained nickel-metal hydride (NiMH) rechargeable batteries that will provide up to 5 hours of operation. Also there is a "AA" alkaline disposable battery pack for NFTI that will provide up to 3 hours of operation. The NFTI has power button and mode button. The two modes of operation are the thermal overlay and video overlay.

The thermal overlay mode displays images that look like a negative picture with the hotter objects appearing brighter than cooler objects. As stated the hotter objects appear brighter than the cooler objects and the temperature is displayed on the screen as well. This mode is very effective for the user when trying to locate hidden hot spots during the overhaul procedure which will be discussed in a later chapter.

The video overlay mode displays images in a visible video format much like a typical camcorder. This mode allows firefighter to see a thermal image superimposed with standard video images in a much better depth perception. This allows the firefighter to easily recognize and identify equipment and objects. In this mode, the temperature off the object in view is displayed on the screen. Additionally, it allows firefighters to see through glass, water, and the ability to read signs and warnings that would not normally be visible in thermal mode alone.

Fire Hoses and Nozzles

The standard Navy fire hose is a double jacketed synthetic fiber with a rubber or similar elastomeric lining. The outer jacket is orange colored pigmentation for easy identification and impregnated to increase wear resistance. Navy fire hoses are available in sizes 1 ½ through 4 inch diameters in 50 foot lengths. The 1 ½, 1 ¾, and 2 ½ inch hoses are used in firefighting and other various equipment such as the portable blowers, in-line eductors and dewatering eductors. For sizes below 4-inch diameter, Navy fire hose has a maximum operating pressure of 270 psi with optimum hose handling characteristics occurring between 90 and 150 psi. Four-inch diameter fire hose are not used for firefighting but they are used with portable educators to discharge flooding water. Portable blowers and in-line eductors will be discussed later in this chapter.

The Navy vari-nozzle is used for firefighting in both fire hoses and AFFF non-collapsible hoses. The vari-nozzle (Figure 3-5) are available in 1 ½ and 2 ½ inches to fit the Navy's three various fire hoses. The spray patterns from this nozzle will vary from straight stream to at least a 90° fog stream pattern. Some nozzles may adjust up to approximately 110° depending on the manufacturer. The spray patterns are changed by rotating the black shroud which surrounds the last 4 to 5 inches of the tip end of the nozzle. Markings are provided to indicate position settings for straight stream, narrow fog (30-45°), and wide fog (90-110°). The flow rate will vary with the pattern used, but the 1 ½ inch nozzle can produce 95 gallons per minute (GPM) and 125 GPM. The 95 GPM vari-nozzle is used on the interior of the ship and the 125 GPM vari-nozzle is used on the weather decks (outside) of the ship. The 2 ½ vari-nozzle is used on the weather decks and produce a flow rate of 250 GPM.

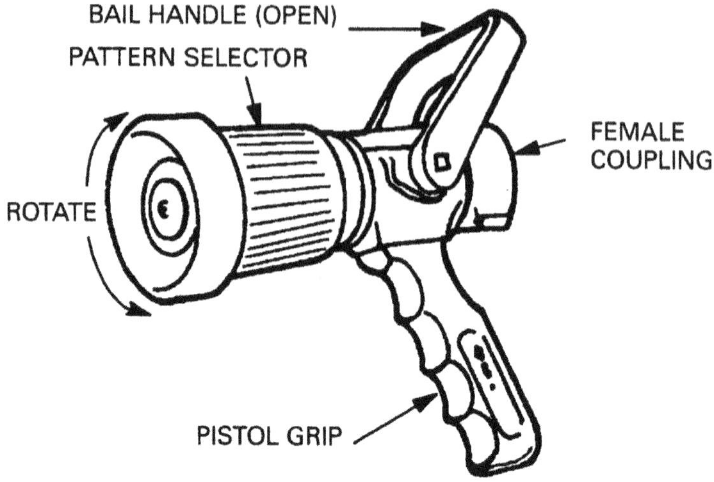

Figure 3-5. Navy 1 ½-inch Vari-nozzle

Portable Afff in-Line Eductors

The portable in-line educator or pick-up tube is used to mix seawater and AFFF concentrate to produce an AFFF solution for combating fires especially Class B fires in machinery rooms or flight decks. The eductor consists of a bronze body with an internal ball that acts as a check valve and flexible pickup tube assembly that is attached to a fire plug and fire hose. The eductor is used in conjunction with a 95 or 125 GPM vari-nozzle. It functions with seawater from the firemain passing through the educator which creates a venturi effect that causes suction in the pickup tube assembly. The educator mixes AFFF concentrate drawn from a 5-gallon can with seawater at approximately 6 percent ratio when the inlet pressure to the eductor is 100 psig. A continuous use will require about 5 gallons of AFFF concentrate per minute.

In-line eductors should be connected directly to fire plugs to minimize inlet pressure reduction due to friction loss. Friction loss downstream of the educator can create sufficient backpressure so the AFFF suction will cease to operate, but seawater will continue to flow. Users of the in-line eductor must limit the hose length downstream of the eductor to three lengths or 50 feet when fighting fires in a horizontal plane or advancing up one deck. A maximum of six lengths of hose (300 feet) may be connected downstream of the eductor whenever the AFFF eductor is rigged on a deck above the affected compartment.

Portable Emergency Pumps

The P-100 portable pump is an engine-driven centrifugal pump assembly that uses diesel or JP-5 fuel. An air-cooled single-cylinder, four-cycle diesel engine rated at 10 horsepower powers the P-100 pump. A mechanical governor is used to control the speed of the engine. It contains a 1.45-gallon fuel tank that will allow up to 2.75 hours of operation. The P-100 is designed to provide 100 GPM of seawater at 83 psi with a suction lift of 20 feet. As a precaution, the P-100 should never be use in poorly ventilated locations because the engine exhaust contains poisonous carbon monoxide gas. The engine exhaust muffler is constructed to receive an exhaust hose that may be attached to route toxic gases to the weather decks.

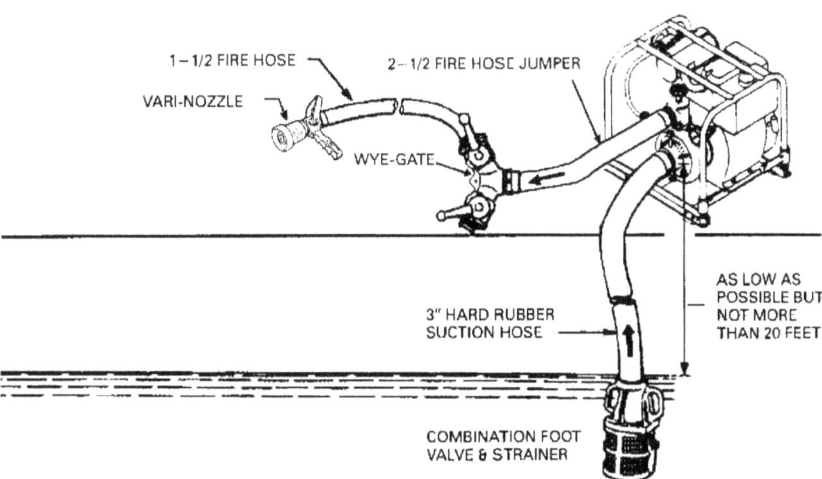

Figure 3-6. P-100 Configured for Firefighting

The P-100 portable pump is an engine-driven centrifugal pump assembly that uses diesel or JP-5 fuel. An air-cooled single-cylinder, four-cycle diesel engine rated at 10 horsepower powers the P-100 pump. A mechanical governor is used to control the speed of the engine. It contains a 1.45-gallon fuel tank that will allow up to 2.75 hours of operation. The P-100 is designed to provide 100 GPM of seawater at 83 psi with a suction lift of 20 feet. The engine exhaust muffler is constructed to receive an exhaust hose. Precautions should be taking when using the P-100 because the engine exhaust contains poisonous carbon monoxide gas. It should never be use in poorly ventilated locations and enclosed spaces without using the approved exhaust hoses that are routed to the weather decks.

Portable Fans and Blowers

Portable fans and blowers are available for desmoking, which is a Navy term for smoke removal (Figure 3-7). The portable fan and blower have their advantages and disadvantages. The electric motor-driven fan is not to be used in explosive environments. The blower is powered by the flow of seawater from the firemain system which requires navy standard fire hoses connected to supply the seawater flow. Both require cleaning, inspection, and maintenance to ensure their reliability which is the responsibility of the fire protection engineer and the electrician by guidance from the planned maintenance system and the manufacturer's technical manuals.

The advantage of these fans or blowers is their ability to recirculate or remove large volumes of air. The Electric "box" fans are convenient and easy to rig but pose risks when operating in explosive atmospheres. Additionally, a box fan depends on electric power for operation which will be a disadvantage whenever the ship loses power. Water-driven blowers pose no threats in explosive atmospheres as long as they are grounded. The disadvantage would be power outage that affects the firemain system, but it can be used with a P-100.

Figure 3-7. Portable Blower and Fan

Portable Ramfan™

The portable Ramfan is one of the primary fans used aboard ships for desmoking or introducing ventilation into a compartment. The Ramfan is driven by water supplied by firemain system or P-100 pump connected to 1 1/2-inch hoses. A water turbine operates the fan blades at approximately 10,000 rotations per minute (rpm), depending on firemain pressure, which will create airflow of 2,000 cubic feet per minute (cfm). The size of the Ramfan is about 18 inches in diameter and weighs about 35 pounds, allowing it to be easily transported. Exhaust hoses can be connected to both sides of the Ramfan. This allows the flow of the smoke or toxic air to be directed over the side of the ship. The operator must take precaution of exhausting explosive gases through the duct because if the Ramfan is not grounded it can create a static electric charge causing an explosive.

Portable Electric Desmoking Fans

The portable electric fan known as the box fan is designed to be used for rapidly desmoking compartments in areas where exhaust ducting is not needed. It produces a tight spiral of air or smoke to prevent recirculation into the area being desmoked. The box fan uses 115 volts of power. The fan will create airflow of 3,200 cfm. The fan has installed hooks used for hanging in doorways to desmoke a space. One of the disadvantages is the flow of the smoke or toxic atmosphere cannot be directed unless it is directly installed near the weather decks. The portable box fan should be inspected for damage before use. Careful inspection of the electrical cord is necessary to prevent shock hazards and the tamper seal on the electric motor must be intact. If this seal is broken, the fan must not be used in any explosive environments. Additionally, the operator must ensure that the screen guards are in place before operation.

This chapter introduced fundamentals of fire protection, fire protection agents, portable fire extinguishers, and portable equipment used during firefighting evolutions. The three elements required for a fire to exist have been identified as well as the classifications of fires. The reader should now be aware of the effects of fire and the different types of gases encountered while fighting fires onboard U.S. Navy ships. No two fires are identical; therefore, the firefighter must rely on their knowledge and training to determine the best extinguishing agent to use when fighting a fire. The use of personal protection gear will keep personnel safe during firefighting evolutions. The firefighting ensemble equipped with the SCBA provides protection for Sailors when fighting fires. This chapter gives a basic overview of what firefighting agents are used onboard ships and when they are used. The next chapter will discuss fire protection systems.

CHAPTER 4

Fire Protection Systems

The keys to success for fighting fires onboard ships depend on the fire protection engineers and crewmembers' ability to operate portable firefighting equipment and emergency repair equipment. Another effective tool is their knowledge and ability to activate all the fire protection systems available onboard Navy ships. A small fire can rapidly spread to fully developed fire that may not be accessible by firefighting teams which can prevent a direct attack on the fire. A compartment can reach the fully developed fire stage very quickly in flammable liquid fires. The key to preventing the fire from fully developing and spreading to other areas of the ship is a quick initial response. That quick response will be dependent on the ability of shipboard personnel to activate installed fire protection systems and use available firefighting equipment. The previous chapters provided general information about firefighting fundamentals, agents, and portable equipment. This chapter will introduce how firefighting agents are used in systems that protect various compartments that contain flammable and combustible materials.

Fire Alarm System

The purpose of fire-alarm system is to detect an occurrence, alert control panels, alert proper authorities, and notify the occupants to take action. A fire alarm system has a number of devices working together to detect and warn people through visual and audio appliances when *smoke, fire, carbon monoxide* or other *emergencies* are present. These alarms may be activated automatically from *smoke detectors, heat detectors* or alarms may be activated through a *manual fire alarm activation* device such as manual call points or pull stations. Naval ships use high temperature alarms and manual operated switches to alert the firefighting and emergency response teams to take proper actions.

High Temperature Alarm System

High temperature alarm systems provide a means of indicating the presence of high temperatures in certain shipboard compartments. The installation consists of two or more thermostats of several types that are located in the following compartments: pilot house, damage control station, interior communications gyro room, magazines (any other compartments where ammunition or critical propellants are stowed), ready service rooms, handling rooms, check-out areas, flammable liquid storerooms, paint mixing and issue rooms, compressed gas stowage compartments, aviation storeroom and other storerooms containing flammable materials and nuclear facilities (other than weapons), cargo holds, cargo spaces and helicopter hangars.

High temperature alarm system for missile launchers provide a means of indicating when the temperature in the missile launcher cells is above or below specified limits. The installation consists of eight sets of thermostatic switches connected to activate audible and visible alarm signals on an alarm switchboard installed in the missile launcher and combat information center (CIC). This will immediately notify the weapons technicians aboard the ships. The weapons technicians can activate magazine sprinkling or deluge systems if necessary to prevent explosions and fires.

Manual Operated Switches

In addition to thermostats, manually-operated lever switches are installed in each conflagration station and OOD station in aircraft carriers, amphibious assault ships and at each exit from helicopter hangars. Audible and visual extension alarm signals are installed in the pilot house, each OOD station, DCC, secondary DCC, and each interior communications room. On aircraft carriers, the special weapons unit compartments are provided with a separate installation for forward and after compartments. A separate installation also is provided for airborne systems support center and the aviation photo interpretation storeroom. Fire alarm pull stations are provided within some machinery spaces on CG-47 and LHD-8 class ships. These fire alarm pull stations are intended to provide a quick initial alarm capability when conditions do not permit a timely verbal report.

Firemain Systems

As mention in the previous chapter, the sea provides an inexhaustible amount of water and the firemain system is filled with seawater pumped from the sea. This system uses seawater pumps called fire pumps to distribute seawater to fireplugs, AFFF systems, sprinkling systems, flushing systems, machinery cooling-water systems, wash-down systems, and other systems as required. The primary function of the firemain system is to supply water to the fireplugs, AFFF systems, and various sprinkling systems, whereas the secondary functions are for systems such as flushing and machinery cooling water. Navy ships use the following types of firemain systems:

6 The single-main system.
7 The horizontal loop system.
8 The vertical offset loop system.
9 The composite system.

The type of firemain system installed depends upon the characteristics and functions of the ship. Small ships generally have a straight-line, single-main system while the large ships have either one of the loop systems or the composite system. The design of the four types of firemain systems are as follows:

1 The single-main firemain system consists of a single piping section that extends fore and aft. This type of firemain system is generally installed near the centerline of the ship, extending in the forward and aft direction of the ship.
2 The horizontal loop firemain system consists of two single cross-connected piping sections that extend forward and aft. The two individual lengths of piping are installed in the same horizontal plane (on the same deck) but are separated athwartships from each other.
3 The vertical offset loop firemain system consists of two single piping sections that are installed fore to aft in an oblique plane. The piping sections are separated both vertically and athwartship and they are connected at the ends to form a loop. The lower section of the firemain is located as low in the ship as practical on one side while the upper section is located on the damage control deck on the opposite side of the ship. Athwartship cross-connect pipes are usually provided at each pump riser.
4 The composite system consists of two service mains installed on the damage control deck and separated athwartship and a bypass main is normally installed on a lower level near the centerline. A bypass section of piping is installed at the lower level near the centerline. Cross-connections are installed alternately between one service piping run and the bypass piping.

Fire Pumps

The Navy standard fire pump (NSTFP) is centrifugal electric motor driven, close coupled pump of titanium construction that is installed in the firemain system to take suction from the sea and distribute the water to primary and secondary systems. The Navy standard fire pump are installed with discharge piping connected to the firemain system and suction piping that connects to a sea chest enclosure that is open to the sea. The fire pumps are provided with suction and discharge cutout valves used for isolation during piping ruptures. The number of fire pumps on Naval ships permit the largest fire demand, all vital continuous loads and the largest backup cooling or 10 percent of all backup cooling demand to be satisfied with only 75 percent of the installed pumps operational. Electric motor driven pumps have normal and alternate independent power sources. The pumps can be remotely controlled from DCC or the ship's damage control deck. They also can be manually operated from the pumps control panel located near the pump. Some fire pumps are equipped with automatic start features that will start a back-up pump once firemain pressure drops below a specific level.

The optimal design goal is to locate at least one fire pump in each fire zone. With this design goal in mind, fire pumps are usually installed below the waterline in separated locations along the length of the ship. The reason they are located below the water line is to provide a minimum positive static head of three (3) feet under all conditions of load and a list of up to fifteen (15) degrees. The fire pump is particularly suited for firemain service because it will deliver a wide range of flow rates as follows:

1. 750 GPM at 125 psi.
2. 750 GPM at 150 psi.
3. 900 GPM at 125 psi.
4. 1000 GPM at 125 psi.
5. 1000 GPM at 150 psi.
6. 1000 GPM at 175 psi.

Fire Hose Stations

A fire hose station is the location of a fireplug valve and the associated firefighting equipment (Figure 4-1). A fire hose station is commonly referred to as either a fire station or fireplug. Fireplugs are valves that branches of the firemain system piping to supply water to fire hose stations throughout the ship. Ships have fire stations with either a 1 ½-inch fireplug valve or a 2 ½-inch fireplug valve. The size of the fire hose station depends on the demand, location, and ships configuration. Fire stations are equipped with a wye gate, fire hoses, vari-nozzles, and an AFFF inline eductor when applicable. The number of fire hoses depends on the demand and location of the fire station. Fire stations can have up to four fire hoses available. Fire stations will have at least one or two connected fire hoses connected to the vari-nozzle and the fire plug. The unattached hoses mounted at a fire station serves two critical purposes:

1. For quick replacement of a damaged hose.
2. For extension of the initial hose length for setting fire boundaries or firefighting.

Figure 4-1. Fire Hose Station

Fire hose stations are installed on ships requiring size 2-1/2 inch fireplugs in quantity and location to permit reaching any weather area of main weather decks and any area on the lower decks from at least two fire stations with 100 feet of hose. On superstructures, fire stations are installed to permit reaching any area from two fire stations with 50 feet of hose. Fire hose stations are installed on ships requiring size 1-1/2 inch fireplugs in quantity and location to permit reaching any main weather deck area or below deck area from at least two fire hose stations with 50 feet of hose. Coverage of all areas of the superstructure is provided from at least two fire hose stations with 50 feet of hose. In addition, fire stations are normally installed at the following locations:

1 Two fire stations in each watertight subdivision containing ammunition transfer areas.
2 At least two fire stations with 150 feet of hose on ships having flight decks in way of maximum width.
3 One fire hose station in the immediate vicinity of spaces where munitions are stowed, handled, or serviced.
4 In machinery spaces, fire stations are installed to provide coverage of all areas with 50 feet of hose from a minimum of one station.
5 Two fire stations are installed to facilitate wash-down of anchor and chain.
6 Fire stations in the vicinity of an external missile launcher are located to permit combatting a fire within the launcher with three hose lines, each 50 feet in length.

All firemain system valve hand wheels and fire plugs are painted red to help shipboard personnel locate the firemain system for firefighting and system isolation whenever there is a firemain piping rupture or leak. In addition to the color coding, each fire plug is identified by a three number locator. These numbers are gives the location such as the deck level, the ship's frame, and the transverse location for example 1-213-2. The first number represents the deck or floor of the ship. The second set of numbers represents the frame number of the ship (from the front of the ship going towards the back) and the third number represents the transverse location of the ship in relations to the centerline. The centerline of the ship is represented with the number 0 and the higher the number the further away from the centerline. An even number will represent the portside (left side of the ship facing forward) and odd number represents the starboard side (right side of the ship facing forward) with higher number outboard of lower numbers. So fire plug 1-213-2 will be located on the 1st deck or floor, 213rd frame, and left side of the ship.

Magazine Sprinkler Systems

Sprinkler systems are used for emergency cooling and firefighting in magazines, ready-service rooms, ammunition, and missile handling areas. A magazine sprinkler system consists of a network of pipes secured to the overhead and connected by a sprinkler system control valve that branches of the ship's firemain system. The pipes are fitted with spray heads or sprinkler-head valves that are arranged so the water can shower all parts of the magazine, ammunition and missile-handling areas. The sprinkler system is designed to wet down all exposed bulkheads at the rate of 2 gallons per minute per square foot and sprinkle the deck area at the rate of 4 gallons per minute per square foot.

Magazine sprinkler systems can completely flood their designated spaces within an hour. To prevent unnecessary flooding, all adjacent compartments are built to be watertight. Upper deck handling and ready-service rooms are equipped with drains that limit the water level to a few inches. The valves that are used to operate various magazine sprinkler systems are as follows:

1 The manual control valve. This valve permits hydraulic operation of the sprinkler valve.
2 The hydraulically operated remote control valve. This diaphragm operated globe type valve is opened by operating pressure acting against the underside of the disk and closed by operating pressure acting on top of the diaphragm. This valve permits the sprinkler

valve to be secured from other stations, whether or not it was manually or automatically actuated.

3 The spring-loaded lift check valve. This spring-loaded, diaphragm operated, lift check valve closes tightly against the reverse flow and opens wide to permit flow in the normal direction. Spring-loaded lift check valves permit the control system to be operated from more than one control station by preventing backflow through the other stations.
4 The hydraulically operated check valve. This valve permits the operating pressure to be vented from the diaphragm chamber of the magazine-sprinkling valve, thereby permitting that valve to close rapidly and completely.
5 Power operated check valve. This piston operated poppet type valve is opened by pressure from the "close" loop of the actuating pressure acting against the piston.

Personnel from the weapons department are responsible for the operation and maintenance of the magazine sprinkler systems. The fire protection must consider what affects their maintenance or repairs on the firemain system will have on the magazine sprinkler systems and other systems. Planning and coordination is used when maintenance and repairs are conducted on firemain system or the magazine sprinkler system. The fire protection engineers must ensure the magazine sprinkler system has a suitable firefighting substitute.

Freshwater Hose Reel System

Freshwater hose reels are installed on aircraft carriers and newer surface ships to combat fires in vital electronic spaces including the inside of electronics cabinets, above false ceilings and under false decks or floors. They are installed adjacent to entrances to vital electronic complexes. Each hose reel is equipped with 50 feet of ¾-inch diameter hose and an adjustable flow, variable pattern nozzle. The fresh water hose reel system is supplied from a fresh water source such as the ship's potable water system. The hose reel service line typically contains an isolation valve, a reduced pressure backflow preventer to prevent contamination of the potable water supply, a dedicated booster pump, a drain, a riser, and piping to a cutout valve at each hose reel's inlet. The activation station is a control panel installed in the vicinity of each hose reel. The control panel consist of two push buttons for powering the system on and off, white power available light, and a green system running light.

AFFF System

AFFF is one of the most commonly used fire protection agents. It is mainly used aboard ships to fight Class B fires and it can be used in conjunction with Purple-K-Powder (PKP). AFFF is delivered through both portable and installed equipment. AFFF systems are installed on naval ships to protect machinery spaces, fueled vehicle stowage spaces, helicopter hangars, landing platforms, refueling stations, flight decks, hangar bays, fuel pump rooms and other compartments or areas where flammable liquid fires are likely to occur (Figure 4-2). Research and previous fire casualties have revealed that these are the places where Class B fires most often occurred. The AFFF systems installed on ships are as follows:

1. AFFF single-speed injection system.
2. AFFF two-speed injection system.
3. Balanced-pressure proportioner (type II).
4. Balanced-pressure proportioner (type III).
5. AFFF eductor system.

Figure 4-2. AFFF System on board a guided missile destroyer (DDG)

AFFF Single-Speed Injection Pump

The AFFF single-speed injection pump is a permanently mounted, positive displacement, electrically driven, sliding-vane type of pump. These pumps are provided in capacities of 12, 27, and 60 GPM. The pump unit is mounted on a steel base consisting of a pump, a motor, and a reduction gear (except 12 GPM that is direct drive) coupled together with flexible couplings. The pump is fitted with an internal relief valve that opens to prevent damage to the pump. The injection pump and the injection station piping are sized to produce a 6 percent concentration by injecting AFFF concentrate into the seawater distribution system. AFFF concentrate is supplied from an AFFF tank. In some installations there is a 1-1/2-inch hose connection between the pump and the pump cutout valve. This hose connection is used to transfer AFFF concentrate to other tanks by connecting hoses between the pump and fill lines of the other tank. AFFF is transferred by manually starting the injection pump that is used to supply the wash-down countermeasure system with AFFF. This action allows the system to be used as a fire-extinguishing system for flight decks, fantail, and helicopter landing flight decks and platforms. Besides wash-down countermeasure systems, injection pumps also supply AFFF to reentry hose reels, well decks, and fueled vehicle stowage decks.

AFFF Two-Speed Injection Pump

The AFFF two-speed injection pump is rated at 27 GPM and 65 GPM. The system is designed to meet the demands for either a low or a high capacity rate for firefighting. The two-speed AFFF pump is mounted on a steel base consisting of a positive displacement pump rated at 175 psi, a motor, and a reducer coupled together with flexible couplings. The pump is designed to inject AFFF concentrate into the seawater supply at a constant flow rate depending on the pressure and demand of firefighting requirements. AFFF concentrations will exceed 6 percent in most cases. The low-speed mode is used for individual AFFF demand from hose reel stations. The high-speed mode is used when firefighting flow rates exceed 250 GPM for hangar bay and deck-edge sprinklers and 450 GPM for bilge sprinklers. The motor on the two-speed pump receives power from a motor controller supplied by a power panel that receives main ship's power from both the ship's service switchboard and the emergency service switchboard. The power panel is equipped with an automatic bus transfer (ABT) to ensure a constant supply of electrical power to the two-speed pump. The motor controller has provisions for both local and remote control. From the local control station, the pump can be activated at either high or low speed. Remote control stations are segregated into high and low demand stations. A high demand station, such as that for a hangar bay sprinkler system, activates the pump at high speed. A low demand station, such as that for a hose reel, activates the pump at low speed. When the system is being secured, the operator can only stop the pump at the local control station.

Balanced-Pressure Proportioner (Type II)

The balanced-pressure proportioner type II system is designed to proportion the correct amount of AFFF concentrate necessary to produce effective AFFF/seawater solution over a wide range of flows and pressures. The system is electrically or manually activated by a solenoid-operated pilot valve (SOPV). The SOPV will vent the operating chamber of the hycheck valve and pressurize the operating chamber of the powertrol valve. The switch assembly of a SOPV via will cause an electrical activation of the pump assembly via the motor controller. The pump assembly is a positive displacement, sliding vane type pump or rotary gear type pump depending on ships design. The pump assembly will pressurize the AFFF concentrate piping to the demand proportioned and balancing valve. Water flow through the proportioner will move the water float towards the large opening of the water sleeve depending on the number of gallons required for firefighting. The AFFF concentrate float is directly controlled by the movement of the water float, thus influencing the amount of AFFF concentrate that enters the water stream.

Water and AFFF concentrate enter the proportioner at the same pressure as indicated for the name balanced pressure-proportioning system. The balanced pressure theory is a direct result of the balancing valve. One sensing line is located downstream of the hycheck valve and another sensing line is located upstream of the AFFF concentrate discharge check valve. They route the water pressure and AFFF concentrate pressure to the operating chambers of the balancing valve. The water-sensing line is piped to the top operating chamber and the

AFFF concentrate-sensing line is piped to the lower operating chamber. Depending on system demands the agent output will increase or decrease through the proportioner by increasing or decreasing the water pressure in the water-operating chamber. When this process occurs, the balancing valve automatically regulates the AFFF concentrate pump discharge pressure to equal the water pressure. The balancing valve will constantly recirculate the AFFF concentrate to the service tank. The amount of AFFF concentrate returned to the service tank through the recirculating piping varies. It depends on the position of the balancing valve which is directly controlled by pressure in the water-operating chamber. Water and AFFF concentrate will be correctly proportioned in the proportioner due to the venturi effect and then discharged the AFFF solution into the distribution piping that leads to the AFFF sprinklers or hose reel.

Balanced-Pressure Proportioner (Type III)

The balanced-pressure proportioner type III was design to replace the type II. The balancing valve concept for the Type III balanced-pressure proportioner is identical to the Type II proportioner. The Type III proportioner houses no internal moving parts and uses the venturi principle to mix the AFFF/seawater solution. An orifice plate controls the volume of AFFF concentrate entering the proportioner. The size of the orifice is determined by the maximum demand for fire protection agent placed on the system design. The water-sensing line connection for the Type III proportioner is piped directly from the proportioner body to the balancing valve. Type III proportioner systems use a positive displacement, sliding vane pump.

Fixed AFFF Eductor Systems

Fixed AFFF eductor systems are installed in ships with limited AFFF services. Each AFFF system eductor is designed to provide AFFF/seawater solution for one specific service such as a hose reel or bilge sprinkling.

An eductor can only provide AFFF/seawater solution within a limited range of flow demand conditions. No AFFF pump is required because the eductor utilizes the high pressure flow from the seawater main to develop the suction or venturi effect to draw out AFFF concentrate from the tank to produce the proportioned AFFF/seawater solution. The AFFF flow is initiated by opening the fire hose valve or sprinkler powertrol valve that is open by a SOPV. The fixed eductor systems are installed onboard Mine Countermeasures (MCM) class ship.

AFFF System Components

Fire protection engineers must have an understanding of the AFFF system's operations and the function of all the system components. Different ships use different AFFF configurations but have similar components. The AFFF single-speed injection pump and the AFFF two-speed injection pump are two types of pumps used with the installed AFFF system. Besides the pumps there are other various components that make up the AFFF system. The primary components and associated equipment of a shipboard AFFF system includes the following:

1 AFFF Eductor.
2 AFFF Tanks.
3 AFFF Valves.
4 AFFF Transfer System.
5 AFFF Transfer Pumps.
6 AFFF Sprinklers.
7 AFFF Hose Reels.

AFFF Eductor

The AFFF eductor may be installed in various systems. An eductor consists of three major components: The nozzle, venturi and body. High pressure fluid enters the body and encounters the nozzle which increases the stream velocity and directs the flow through the venturi throat. The low (suction) pressure AFFF concentrate stream is introduced into the high pressure fluid in the venturi throat. This is where the AFFF concentrate and seawater is mixed. The AFFF mix then travels through the diffuser outlet of the main body and exits the eductor to the AFFF sprinkler system or AFFF hose reel.

AFFF Tanks

AFFF concentrate is stored in service tanks of 50 to 2,000-gallon capacity and storage/transfer tanks up to 3,500-gallon capacity. The tanks are rectangular or cylindrical in shape and they are fabricated out of copper-nickel (90/10) or corrosion-resistant steel. Each service tank is located inside the AFFF station and they are fitted with a vent, drain connection, fill connection, liquid level indicator, recirculating line, and an accessible manhole cover to enter the tank for maintenance. The vent prevents excessive buildup of pressure within the tank during storage and prevents a vacuum when the system is in operation.

AFFF Valves

The AFFF system requires a variety of valves with different functions. These valves help with creating the AFFF mixture, directing the flow to the needed location, or preventing back flow. These valves included in the AFFF systems and their functions are as follows:

1. Powertrol valve.
2. Powertrol valve with test connection.
3. Powercheck valve.
4. Hytrol valve.
5. Hycheck valve.
6. Solenoid-operated pilot valve (SOPV).
7. Balancing valve.

Powertrol Valve

The Powertrol valve is a diaphragm type, normally closed, seawater pressure operated control valve that allows the flow of AFFF/seawater solution through the distribution system, or controls seawater flow on flight deck injection systems. The Powertrol is held closed by spring pressure until the valve is actuated when the control line is pressurized using firemain pressure directed by the SOPV. This control line pressure is exerted on the bottom of the valve's diaphragm and the valve is forced open against the spring allowing the flow of AFFF/seawater solution.

Powertrol Valve with Test Connection

The powertrol valve with test connection is a diaphragm-type, hydraulically operated, globe control. This valve is normally used as a sprinkler group control valve but also it used by fire protection engineers to conduct maintenance and system testing. This valve is essential because it would be impossible to test a sprinkler group unless the fluid flow could be diverted before it is discharged through the sprinkler heads. To perform a system test, the test connection cap is removed from the bottom of the valve and a test fitting inserted. The test fitting has an O ring gasket to provide a seal between the fitting and the valve seat. A drain hose or fire hose is connected to the fitting and the discharge end of the hose is place in a suitable location. When the operating chamber is pressurized, the valve opens to divert seawater or AFFF solution through the test fitting and out the drain hose. The valve is successfully tested and the sprinkler groups remained dry.

Powercheck Valve

The powercheck valve is a diaphragm type, normally closed, seawater pressure-operated control valve. This valve allows flow of AFFF from the pump to be mixed with seawater and protects the AFFF tank from seawater contamination or dilution. The powercheck is essentially a powertrol valve that has a lift-check feature built into it. When there is no pressure on the control line, the upper valve spring forces the valve closed and the lift check feature is inoperative. When there is firemain pressure on the control line, this pressure acts on the bottom of the diaphragm and opens the valve against the upper valve spring. The flow of AFFF pushes the disk and lower stem upwards which allows AFFF to flow. If the back pressure downstream of the valve exceeds the pressure on the upstream side of the valve, the valve disc holder and stem will slide to the closed position, preventing any backflow through the valve. On balanced pressure proportioning systems, the powercheck does not have a lower spring.

Hytrol Valve

The hytrol valve is a diaphragm type, fail open, seawater pressure-operated control valve which controls the flow of AFFF solution to systems. When in ready status, the top of the diaphragm is subject to spring pressure and control line pressure. When the control line pressure is removed, the firemain pressure overcomes the spring pressure and opens the valve. When an SOPV directs firemain pressure to the control line, the pressure on top of the diaphragm balances the firemain pressure on the bottom of the diaphragm and allows the spring to close the valve.

Hycheck Valve

The hycheck valve is a diaphragm type, fail open, seawater pressure-operated control valve which allows the flow of seawater from the firemain system to be mixed with AFFF concentrate. The hycheck is equipped with a sliding lift check feature like the powercheck valve. When the AFFF system is in standby, the hycheck is held closed by firemain pressure on top of the diaphragm. The master SOPV maintains the firemain pressure on the diaphragm and when firemain pressure on the diaphragm is relieved by the master SOPV, the hycheck valve will open to allow flow. If foam demand stops and the AFFF/seawater solution pressure equals or exceeds the firemain pressure, the lower spring closes the valve disk. This prevents the AFFF pump from pumping concentrate back into the firemain system.

Solenoid-Operated Pilot Valve (SOPV)

SOPVs are electrically operated pilot valves that control the activation of many AFFF fire protection systems. Master and service SOPVs are installed with four control line ports. One port is always connected to firemain pressure (supply) and a second port is the valve drain (which is piped to discharge within the coaming of the AFFF station). The other two control ports are connected by control lines to diaphragm-operated control valves on master SOPVs and the service SOPVs have one control port plugged.

Balancing Valve

The balancing valve automatically proportions the correct amount of AFFF concentrate with seawater. The balancing valve is a diaphragm-actuated control valve that responds to pressure changes between the AFFF concentrate supply line and the firemain. Two sensing lines are attached to the balancing valve, one to monitor the pressure in the AFFF concentrate piping, and one to monitor the firemain pressure. The pressure differential between these lines moves the diaphragm in the control valve. As the AFFF/seawater flow increases, the firemain sensing line pressure drops and the control valve adjusts by forcing more AFFF concentrate into the proportioner.

AFFF Transfer System

AFFF generating stations use large volumes of AFFF concentrate during fire-fighting evolutions. The service tank alone may not contain enough AFFF concentrate to combat a conflagration-type fire. Transfer capabilities are available to replenish the AFFF concentrate service tanks. The installed system consists of a reserve transfer pump (positive displacement, sliding vane, or centrifugal), reserve storage tanks, associated piping, and valves. The transfer system can deliver AFFF concentrate to service tanks through the transfer main piping. The transfer main consists of a large pipe with smaller branch connections interconnecting the AFFF service and storage tanks. This feature gives the on-station concentrate pump the capability of delivering AFFF concentrate into the transfer main. Once the transfer main is pressurized, either by the reserve pump or by the on-station pump, all AFFF generating station service tanks can be replenished. On-station pumps used in conjunction with jumper hoses and hose connection valves may be used to transfer AFFF concentrate. Some ships do not have a transfer system but can utilize another AFFF station service tank through aligned cross-connected piping and valves. Some ships can replenish the service tanks with 55-gallon containers located near the generating station. They do this with an installed hand-operated pump or air-regulated transfer system. The air-regulated transfer system may be used to replenish reserve storage tanks. Ships may replenish service tanks or storage tanks by manually dumping AFFF concentrate from 5-gallon containers through a fill connection.

AFFF Transfer Pumps

The AFFF transfer system may also use a pump to transfer AFFF. The AFFF transfer pump is a permanently mounted, single-speed, centrifugal type, electrically driven pump. These pumps are provided in 360-GPM capacity. The transfer pump moves AFFF concentrate through the AFFF fill-and-transfer subsystem to all AFFF station service tanks on a selective basis.

AFFF Sprinklers

AFFF sprinkler systems are a convenient and quick method for the firefighting teams to apply AFFF solution to large areas of burning fuel. The system consists of a large header pipe with branches of smaller piping with attached sprinkler heads. A sprinkler group control valve (powertrol or hytrol with test connection) will control the discharge flow to the sprinkler heads. A SOPV or a manual control valve may actuate the group control valve. Some sprinkler systems are activated by a manually operated cutout valve. Activation controls for the group control valves depends on the sprinkler system installation and type of ship, but usually the activation controls may be located in primary flight control, the navigational bridge, the helicopter (helo) control, a conflagration station, locally at the AFFF generating station, and at various locations throughout the ship. An AFFF sprinkler system is a subsystem of AFFF generating systems. Some of the different types of sprinkler systems aboard naval ships are as follows:

1 The bilge sprinkling system is located in the main and auxiliary machinery spaces with the sprinkler heads installed below the lower level deck plates. Overhead sprinkling is installed in the overhead of helicopter and hangar bays, well decks, vehicle cargo holds, and fuel pump rooms. Some diesel-powered ships have the overhead sprinkler system installed in the main machinery space.
2 The flush-deck system uses the countermeasure wash-down flush-deck nozzles to discharge AFFF/water solution during flight deck and helicopter deck fires. This capability is currently available to all aircraft carriers, helicopter carriers, and some auxiliary and combatant ships.
3 The deck-edge sprinkler sprays AFFF/water solution over the flight deck of aircraft carriers and helicopter carriers. The system consists of spray nozzles that are positioned at the deck-edge combing of the port and starboard sides on helicopter and flight decks. The nozzles project the AFFF/water solution across the deck in an arc pattern to spray over the top of the burning fuel and aircraft.

AFFF Hose Reels

The AFFF hose reels consist of a steel wheel drum with an attached 1 ½-inch rubber non-collapsible hose that is manufactured in various lengths (Figure 4-3). The AFFF hose reels provide a swivel connection to the piping which permits the hoses to be uncoiled from the reels while still connected to the piping. A hand crank system is incorporated into the reels to facilitate rewinding of the hoses. AFFF hose reels are provided for entry into machinery spaces and to protect aviation facilities. AFFF enters the hose reel through a swivel joint and flows into an elbow located inside the drum. AFFF then flows through the outlet elbow which protrudes from an opening in the drum. The hose is connected to the outlet elbow and then wrapped around the wheel drum for storage. On some reels the outlet elbow is a swivel or reversible type. The AFFF Hose reels use vari-nozzles to effectively provide proper patterns to fight the fire.

Figure 4-3. AFFF Hose Reel

AFFF System Maintenance and Testing Equipment

It is imperative that all AFFF system equipment and components be in functional condition when an emergency arises. Current Planned Maintenance System requirements and technical manuals provide step-by-step procedures for maintaining shipboard AFFF systems and components. Specific maintenance requirements and periodicities are provided on Maintenance Requirement Cards. AFFF concentrate and AFFF/seawater solution must be tested periodically to ensure that the firefighting teams have an effective agent to combat Class B fires. To accomplish the test, fire protection engineers must ensure that the AFFF/seawater has the proper mixture ratio and that the tank is not contaminated. They used testing equipment to ensure proper AFFF/seawater ratio and contamination. The equipment that is used includes the hand refractometer and the quantab chloride titrator strip.

Refractometer

The hand refractometer gives accurate readings of total dissolved solids in aqueous solutions. If an AFFF generating system is tested according to maintenance procedures. The refractometer reading indicates the percent of solids present in the solution. To get proper readings, samples must be drawn from the same water source and AFFF concentrate service tank that were used to generate the AFFF/water solution. For example, if a ship has 20 AFFF generating systems, then there must be 20 AFFF concentrate samples and 20 AFFF/water solution samples. The refractometer will be used to determine the percent of solids present in the aqueous solution samples. Once all the readings have been taken, the fire protection engineer can determine the percent of AFFF concentrate that is being proportioned with water by using the following formula:

RS - RW = A RT - RW = B

A **x 100 =** ***Percent of AFFF concentrate B***

RS = the hose sample (AFFF/water solution)

RT = the service tank sample (AFFF concentrate)

RW = the water sample

Quantab Chloride Titrator

Quantab chloride titrator strips are used to measure salt (chloride) in aqueous solutions. Seawater contains approximately 20,000 ppm of chloride. The allowable limit for chloride contamination of AFFF concentrate is 2,000 ppm, which equates to a 10 percent contamination. There is a caution that must be taken for all approved AFFF concentrates have been subjected to 10 percent seawater contamination tests and have passed corrosion tests for metals approved in AFFF generating systems. Contamination above the 10 percent limit causes two problems: (1) AFFF generating system components will corrode and (2) Improper AFFF/water solutions result in an ineffective firefighting agent.

If contamination exceeds 2,000 ppm, fire protection engineers must identify the source of contamination and correct it before dumping the contents of the AFFF concentrate tank. All AFFF concentrate components must be clean before replenishing the service tank. Once the service tank has been replenished, all AFFF testing procedures shall continue according to the planned maintenance requirements. There is a warning that must be adhere to when conducting testing and maintenance on AFFF systems because of the potential for hydrogen sulfide formation exists in stagnant premixed solutions of seawater and AFFF concentrate. Personnel are at risk in the vicinity of the point of discharge during AFFF system operation or maintenance, particularly in poorly ventilated spaces. Fire protection engineers must take precautions by being equipped with atmosphere testing devices along with ready available breathing apparatus.

Installed Carbon Dioxide (CO_2) Floodig Systems

As mention in the previous chapter, CO_2 is a colorless, odorless gas that is naturally present in the atmosphere at an average concentration of 0.03 percent. It extinguishes fires by reducing the concentration of oxygen in the air to the point where combustion stops. CO_2 concentrations ranging from 30 to 70 percent are required to extinguish fires. CO_2

systems are installed on naval ships to provide a dependable means to flood or partially flood certain areas that present unusual fire hazards. An installed CO_2 extinguishing system has one or more 50-pound cylinders. The cylinders may be installed as a single unit or in a bank of cylinders of two or more. The number of cylinders in a bank is dependent upon the specific requirements of the area protected. The 50-pound CO_2 cylinders are essentially the same as the 15-pound portable CO_2 extinguisher except for the size and actuation mechanisms. The CO_2 installed systems are being replaced with HFP on newer ships.

The CO_2 flooding system is used for spaces that are not normally occupied by personnel. The system consists of one or more cylinders that are connected by piping from the valve outlets to a manifold. Fixed piping leads from the manifold to various nozzles installed throughout the compartment. Cables run from the valve control mechanisms on the CO_2 cylinders to the pull boxes located outside the compartment. The CO_2 cylinders maybe located outside of the protected compartment depending on the construction of the system. Once the CO_2 is needed, the operator will break the pull box glass and pull the handle of the cable leading to the valve control mechanisms on the cylinders releasing the CO_2 from the cylinder.

There are usually one or two valve control devices in a CO_2 flooding system. The number of valve control devices provided will depend on the number of cylinders in the bank. The remaining cylinders in the bank are provided with pressure-actuated discharge heads. These heads open automatically when pressure from the controlled cylinders enters the discharge head outlet. Once the system has been operated, the pressure operated switches, mounted outside the protected area near the compartment's access, will activate. The switches are operated by CO_2 pressure and they are utilized to shut down ventilation, activate flashing red lights, and sound an alarm for the affected spaces.

The CO_2 flooding systems discharge time delay is set for 30-seconds to provide adequate time for personnel to evacuate the space. CO_2 enters the inlet connection on the body and flows through the metering tube assembly to the accumulator. As the pressure increases in the accumulator, the CO_2 also begins to pressurize the chamber at the top side of the piston. When sufficient pressure is attained, the piston depresses the main check assembly, allowing the CO_2 to pass straight through the body to allow it to flow to where it is needed.

Personnel must take precautions before activating a CO_2 flooding system. Before activating an installed CO_2 system, personnel must ensure that all openings in the compartment are closed and the ventilation system for the compartment is secured. These precautions are necessary to prevent the loss of CO_2. If the CO_2 has been exhausted, immediately check for openings and ventilation running and once they have been closed and stop then secondary may be utilized (only if installed).

All Navy CO_2 cylinders are held under a pressure of 850 psi at 70 °F. Any increase in temperature increases the pressure and any decrease in temperature decreases the pressure. Since pressure of liquefied CO_2 builds up rapidly as the temperature rises, the following procedures are utilized to avoid the danger of explosion. The rated capacity of the cylinders is never more than 60 percent of the liquid volume capacity of the cylinder, or 68 percent for portable CO_2 fire extinguishers. CO_2 cylinders are hydrostatically tested to a minimum of 3,000 psi. The CO_2 cylinders are fitted with a bursting disk designed to relieve from 2650 to 3000 psi. Therefore, these cylinders can be installed in spaces of temperatures up to 135 °F with a full 50-pound charge.

Carbon Dioxide (CO_2) Hose and Reel Systems

CO_2 hose reels are used to extinguish electrical fires in large switchboards and generators in nuclear and electric driven ships. The CO_2 hose-and-reel installation consists of two cylinders, a length of special CO_2 hose coiled on a reel, and a horn-shaped nonconductive nozzle equipped with a second control valve (Figure 4-4). When the hose and reel are both installed near the normal access, each of the two cylinders may be actuated individually. The cylinders may not be located near the hose reel and in this case a manual pull boxes are located near the hose reel to activate the system. The high pressure hose (minimum 5,000 psi bursting pressure) is stowed on a permanently mounted trunnion type reel. Piping from the cylinder is linked into the CO_2 connection in the axle of the reel. A gland and packing are provided at this point to seal the connection against leakage as the reel is rotated to unwind or rewind the hose. A swivel nut is used to attach the hose to the reel so that the hose can be readily replaced without having to twist the whole length of hose.

Figure 4-4. CO2 Hose and Reel System

Halon And Heptafluoropropane (HFP) System

As mention in the previous chapter, halon is a halogenated hydrocarbon used for fire protection. Halon 1301 is the type of halon used in installed systems onboard Navy ships (Figure 4-5). Halon 1301 is preferred over halon 1211 because halon 1211 has a low volatility combined with a high liquid density which permits the agent to be sprayed as a liquid. Halon 1301 is super pressurized with nitrogen and stored in gas cylinders as a liquid. Halon 1301 is used in machinery spaces and flammable liquid storage and issue rooms because of its effectiveness against those fires requiring total gas flooding, low toxicity, lack of agent-induced damage, and a relatively low system space and weight impact. This agent is suitable for flammable gas, liquid, and typical solid combustible fires. It extinguishes fires in enclosed spaces by employing the principle of gas-phase catalytic interruption of combustion reactions.

When halon is released, it vaporizes to a colorless, odorless gas with a density of approximately five times that of air. Halon 1211 and halon 1301 are severe ozone depleting substances. These agents should be used only against actual fires. Because of their ozone depleting properties, newer Navy ships have been using alternative fire extinguishing systems for halon. HFP is a fire extinguishing agent that was developed as a drop-in substitute. HFP is a colorless, odorless and electrically non-conducting gas. HFP is clean agent that leaves no residue. HFP is stored in steel containers at 600 PSIG at 70 ° F (41 bars at 21 ° C), as a liquefied compressed gas, with nitrogen added to improve the discharge characteristics. HFP is a different agent than halon but uses the same system components. For this section halon 1301 and HFP will be discussed together.

Figure 4-5. Halon Cylinder Bank

Halon 1301 or HFP systems may be installed in Main Machinery Rooms, Fire Rooms, Engine Rooms, Auxiliary Machinery Rooms, Fuel Pump Rooms, Ship Service or Emergency Generator rooms, Auxiliary Boiler Rooms, Main Propulsion or Generator Engine Modules, Helicopter Recovery Assist, Securing, and Traversing (RAST) Machinery Rooms, Tactical Towed Array Sonar (TACTAS) Handling Rooms, and in spaces where flammable/combustible liquids are stored or issued. In Aircraft Carriers, MOGAS (motor gasoline) powered bomb hoist storerooms may be protected by halon 1301 or HFP. Halon or HFP systems utilize one or more cylinders containing halon 1301 or HFP in a liquid form.

System Components

The function of the system is to extinguish fires which are beyond the scope of manual firefighting equipment and where abandonment of the compartments necessary. The halon 1301 or HFP normally have a primary and reserve system independent of each other. The number and size of cylinders will vary depending on the size of the compartment protected. HFP system requirements are very similar to halon 1301 and their components include the following:

1 Halon 1301 or HFP cylinders.
2 Manual 5-pound CO_2 actuators.
3 CO_2 Vent Fittings.
4 CO_2 Actuation Piping.
5 Flexible Discharge Hoses.
6 Check Valve.
7 Time-Delay Device.
8 Time-Delay Device Bypass Valve.
9 Pressure Switches, Electrically Operated Alarms, and Indicators.
10 Halon or HFP Discharge Piping and Nozzles.
11 In-line Filter.

Halon and HFP Cylinders

Both the halon 1301 and HFP cylinder assembly consists of a high-pressure cylinder, a siphon tube that reaches to within approximately 1-1/2 inches of the bottom of the cylinder, and a pressure operated valve with a pressure gage. The cylinders contain either halon or HFP in its liquid form pressurized with nitrogen at 600 to 675 psi at 70°F (21.1° C). Halon 1301 cylinders are found in 10, 15, 60, 95, and 125 pound capacities. Since HFP is denser than halon, its cylinders are found in capacities of 21, 62, 98, and 129 pounds. Halon and HFP cylinders can be distinguished by the names of the agents painted vertically on the cylinders. Additionally, halon cylinders will have a large white stripe over a gray stripe located near the top of the cylinder. HFP cylinders will have a gray stripe over a black stripe.

Manual 5-pound CO_2 Actuators

The manual actuation cylinder is the principle means of actuating the halon 1301 or HFP system (Figure 4-6). The actuation cylinder is fitted with a manual valve and charged with 5 pounds of CO_2. The CO_2 acts as the pneumatic medium for the operation of the halon or HFP cylinder valves. The CO_2 is held in the cylinder by the valve seat and by the rupture disk. To release the CO_2, the tamper seal is broken, the pin is removed, and the manual control lever is operated. The cam action of the manual lever depresses the actuating plunger which opens the valve seat discharging the CO_2. Once the CO_2 is discharged, the alarms will sound, the ventilation will shut down, and the halon or HFP will discharge into the compartment. The 5-pound CO_2 actuation cylinders may be found marked specifically for the halon and HFP systems, but these are functionally identical and can be used in either type of system.

CO_2 Vent Fitting

The CO_2 vent fitting is a pipe plug with a small 1/32-inch hole that allows the venting of pressure in the actuation piping. The vent plug is installed at the end of the actuation tubing near the halon or HFP cylinders. This vent fitting will allow the excess CO2 pressure to bleed out the actuation piping after operation. Once CO2 pressure has been released, the system components can then be safely disconnected to restore the halon or HFP system back to normal operations.

CO_2 Actuation Lines

The actuation lines are used to carry the CO_2 from the manual actuation bottle to the halon or HFP. The actuation lines consist of 1/4-inch copper-nickel tubing. The actuation tubing is bent in several loops ranging from 1 1/4- inch to 4 inch. Because of size specification, the actuation tubing may not be locally fabricated and replaced by ship personnel. The fire protection engineers must replace the actuation tubing that is sent from the manufacture.

Figure 4-6. Halon 1301 Actuation Station

Flexible Discharge Hoses

The 1-1/2 inch all-metal hoses connect halon or HFP cylinders to the discharge piping. This hose provides a convenient disconnect point for removing halon or HFP cylinders for charging or maintenance. The flexible hoses are of all steel construction that has interwoven metal braids and must be replaced whenever multiple braids are broken or if the hose has a bulge in it. At no time this flexible discharge hose shall be painted.

Check Valve

Check valves in the CO_2 actuation piping allow the CO_2 to flow in one direction. This prevents backup of gas pressure in the opposite direction. Once the halon or HFP has been activated the check valves must be cleaned and re-installed. An engraved arrow is located on the side of the check valve to prevent the check valve from being installed backwards.

Time-Delay Device

The time-delay device is installed in the CO_2 actuation system between the manual 5-pound CO_2 actuation cylinders and the halon or HFP cylinders. The time delay gives personnel time to evacuate, secure accesses to the compartment, and time for ventilation systems to shut down before the agent is released into the affected compartment. The time-delay device will provide either a 30 or 60 second delay of halon or HFP discharge after system activation. The 30 second time-delay device are installed in small compartments that are not normally occupied by personnel, whereas the 60 second time-delay device are installed in larger compartments that are occupied by personnel.

Time-Delay Device Bypass Valve

The time-delay device bypass valve is a ball valve connected to the inlet and outlet piping of the time delay device. The valve handle is painted gray with a white stripe for identification. When manually opened, the valve allows the CO_2 to bypass the time delay device and actuate the halon or HFP. The time-delay device bypass valve shall always be in the closed position and attached with a tamper seal. Once primary halon or HFP has fail to operate or the reserve system needs to be actuated, the time-delay device bypass valve can be immediately open after actuation.

Pressure Switches, Electrically Operated Alarms, and Indicators

The pressure operated switch consists of a two-position, three-pole contactor, mounted in a spray-tight enclosure. A pressure operated cylinder actuator moves the switch toggle from its normal (knob down) position to its actuated (knob up) position when pressure is applied to the cylinder. The CO_2 pressure pushes up the switch actuator piston and stem which is connected to the switch lever operating assembly. There are three pressure switches that are actuated when halon or HFP is actuated. The first two pressure switches that are actuated are the audible/visual alarm switch and the ventilation shutdown switch. The last switch to actuate is the discharge switch. This switch will actuate once halon or HFP is released into the affected compartment.

Each halon or HFP actuation station has an indication panel of system status. The panel will always have a white power available light. The halon or HFP system activated light will illuminate orange once the system has been activated. The discharge indication light will illuminate red once the halon or HFP agent has been discharged. These pressure switches and indication lights have circuits that will activate alarms in Damage Control Central, the ship's pilot house, and various controlling areas throughout the ship.

Halon or HFP Discharge Piping and Nozzles

The halon 1301 or HFP system consists of one or more cylinders piped to discharge nozzles located at the overhead of each level. The discharge piping is use to disperse the halon or HFP agent throughout the protected compartment. The flexible hose connects the halon or HFP cylinder to the discharge piping. In spaces with obstructions or very high overheads, nozzles may also be placed at intermediate levels. Cylinders may be arranged in banks with common piping leading to multiple nozzles, or may be of a modular type with cylinders located throughout the protected compartment with separate piping from each cylinder to a discharge nozzle. Nozzles control the rate of discharge from the system pipelines, thereby regulating the distribution of halon or HFP throughout the compartment. The orifice size of the nozzle is determined by the volume of the portion of the compartment being protected and the pressure available at the nozzle.

In-line Filter

The Filters are provided upstream of each time-delay device to collect any contamination that may exist in the CO_2 actuation piping. This filter protects the small metering orifice in the time-delay device. This filter is 10 microns and it is installed in the CO_2 actuation system upstream of the time delay device. This filter improves system reliability and it should be cleaned after the halon or HFP system gets restored.

Location

The location for halon or HFP cylinders can be found on the inside of the protected compartment or they may be located outside the compartment or in a designated halon or HFP cylinder room. Halon or HFP systems placed in machinery spaces such as Main Machinery Rooms, Firerooms, Engine Rooms, and Auxiliary Machinery Rooms will have 60-second time delays. In compartments other than Machinery Spaces, halon or HFP systems usually have a 30-second time delay and may only have a primary halon or HFP system. Engine enclosures or modules have a 30-second time delay for both primary and reserve halon systems.

Capabilities

Each system is designed so a single discharge of halon 1301 or HFP provides a concentration of 5 to 7 percent by volume of air throughout the protected space. Sufficient halon or HFP is required so the Concentration will remain at a minimum of 5 percent for 15 minutes. Some halon or HFP protected spaces have a duplicate reserve halon or HFP system to supplement the primary one. Each halon or HFP fixed-flooding system is designed to discharge the agent into the protected compartment within 10 seconds following the start of the discharge.

System Actuation, Features, and Operation

Each system is usually provided with more than one CO_2 actuator station. The actuators can be installed either inside or outside the space. Features of the system include automatic ventilation shutdown, actuation of local and remote alarms, manual time delay bypass, and halon or HFP discharge indicator light. Normal operation of the halon system may be accomplished by performing the following actions:

1 Break the glass or open the enclosure at remote actuating stations. Remove the safety pin, which is secured by a tamper seal.

2 Fully operate the discharge lever and secure it in the OPERATE position. The released carbon dioxide will immediately actuate two pressure switches. One pressure switch operates lights and horns (or bells) within the space, and a bell and amber system actuated light outside the compartment at actuating stations and accesses. The other pressure switch will initiate shut down of ventilation fans and operate any installed vent closures.

3 If alarms do not operate, or ventilation does not shut off, pull out the reset/actuation knob on the associated pressure switch. If operation still does not occur, manually shut off ventilation systems, and pass the word to evacuate the space.

4 After the time delay operates, the carbon dioxide pressure will operate the halon or HFP cylinder valves to discharge halon to its associated nozzles. A third pressure switch downstream of the time delay device will then actuate a red light indicating halon or HFP discharge.

5 In the event the timing of the time delay device exceeds 70 seconds (for a 60-second device), or 35 seconds (for a 30-second device), the time delay should be bypassed by opening the time delay bypass valve. **WARNING:** The time delay bypass valve should not be operated until after the full delay time of 30 or 60 seconds has passed.

Additional features include automatic ventilation shutdown, actuation of local and remote predischarge alarms, manual time delay bypass, automatic ventilation closures (if installed), and halon or HFP red discharged indicator light. An AFFF bilge sprinkling system normally supplements halon 1301 systems in machinery spaces and pump rooms. The AFFF bilge sprinkling system, where installed, should be actuated at the same time as the halon system. AFFF bilge sprinkling systems are not installed if the bilge is too shallow.

Water Mist Systems

Water mist is a fire protection agent which replaces halon 1301 in new ship designs such as the LPD 17 and littoral combat ships (LCS). Water mist for machinery spaces is a total-space fire extinguishing system which discharges high-pressure (approximately 1000 psi) fresh water as a fine mist from nozzles located in all levels except the bilge. High-pressure water mist is effective at suppressing oil pool fires, oil spray fires, and Class A fires even if the fire is obstructed from the nozzles. Water mist may not totally extinguish deep-seated Class A fires, but will knockdown open flames to a smoldering state and prevent spread of a Class A fire. Actuation of the water mist system will automatically shut down ventilation in the affected space, but does not include a 30 second or 60 second actuation time delay needed for ventilation air-flow stoppage as with halon 1301 or HFP systems. Water mist may cause short circuits in energized electrical equipment.

The high pressure water mist system consists of dedicated fresh water tanks, pumps, distribution piping, nozzle grids in each level of each protected space, and associated controls. The system is designed to provide a minimum of 15 minutes of continuous operation of water at a flow rate of 0.003 GPM/ft3 to 0.0045 GPM/ft3 for the protected space. The application density or volumetric flow rate is based on the gross volume bounded by the compartment's forward and after bulkheads, the longitudinal bulkheads or shell as applicable, the nominal deck plate and overhead. Pumps and tanks are properly sized to provide this volumetric flow rate to all levels of the largest protected space. Pumps and piping are fitted so that the inlet pressure to each nozzle is between 900 lb/in2 and 1100 lb/in2 with a desired value of 1000 lb/in2.

The system architecture is a single main located centrally in the ship for protection against damage from attacks on the ship. Although the system is a single main, there are dual flow pathways available for each main and auxiliary space. The valve actuators are fire resistant with an overwrap of insulating material because many of the isolation and distribution valves are in spaces where they could possibly be exposed to fire. The normal condition of the main is always filled and unpressurized. The wet status of the system and the location of the distribution piping reduce lag time between system actuation and the beginning of water mist flow. All electrically operated components are provided with normal and backup power supplies. Two pumping stations are provided for redundancy and survivability. If the system is undamaged, multiple spaces can be provided with flow simultaneously up to the pumping capacity of the system. The normal system controls will operate the correct pumps for the flow demand of the affected compartment. The operator can override the control system. If too many spaces are aligned for mist discharge, the residual pressure in the system will drop below the minimum required nozzle pressure and the system will not be effective at controlling the fire. If too many pumps are aligned for mist discharge, the residual pressure in the system will cause the relief valve to lift. Both of these scenarios will generate alarms.

Proper system operation demands clean water at high pressure. High pressure is required for satisfactory operation of the nozzles so the number of potential discharge openings from the system has been minimized and most drains and vents are fitted with blanks and valves. An orifice at each nozzle will restrict the flow if a nozzle is missing or broken which will preserve pressure for the remaining nozzles. Strainers are installed throughout the system to ensure the nozzles do not get clogged.

On most ships, each storage tank is fitted to provide flow to the largest protected compartment for at least 15 minutes. On LCS, smaller tanks are provided to help reduce the weight. These ships refill operations are started automatically when the water mist system is activated. The tanks are filled from the ship's potable water system with an emergency hose connection fill capability from the firemain. The tanks can be refilled even while water mist pumps are operating or the second pumping station can be placed online when the first tank is depleted. The gross tank volume includes allowances for internal structure, the estimated depth of water in the tank at loss of pump suction, and the maximum conditions of list and trims which are specified for the particular ship. For maintenance and inspection, the tanks can be dewatered by the drainage system.

The pumps are motor driven quintuplex reciprocating plunger types which simply means there are five cylinders. The positive displacement plunger pumps are capable of producing the high pressure needed at the nozzle and the five cylinders help reduce the peak pulsation pressure that would occur with fewer cylinders. The pumps have an inherent "check valve" capability so when one pump station is operating water will not be forced through the idle pump into its tank.

The valves are electric motor-operated quarter-turn ball valves for system isolation and for aligning source water to the individual spaces when required. Valves in the distribution main are designated as isolation valves. Valves in the distribution branches are designated as primary and secondary distribution valves. Each protected compartment is provided with an isolation valve forward or aft of the compartment use to isolate that portion of the distribution main when required. If the adjacent compartment is an occupied machinery space, these valves are provided with hand wheels or operating levers located in the adjacent for backup manual operation of the motor-operated valve.

The nozzles produce the water mist due to the shearing action of high pressure water passing through the internal passages and orifices. The nozzle assembly incorporates seven tips for omnidirectional spray. There is no logistics support for the individual elements of the nozzle;

therefore, it is intended to be replaced as an assembly. If a nozzle is ever missing, an orifice upstream of each nozzle will restrict flow and preserve sufficient residual pressure for the remaining nozzles.

The system is operable from the ship's damage control system located in DCC, control panels on the main (DC) deck, and from local controls at the various components. Sensors and control logic will alert the operator whenever a variety of faults occur, such as pump and valve failures, or mismatches between pump capacity and system flow demand. When there is a fault or an expected action fails to occur, the operator must align the system to provide flow to where it is needed. On LHD 8 class, the machinery control system will take automatic action to start a backup pump or open a secondary distribution valve, but the operator may also take manual control.

On LHD 8, the machinery control system monitors the water mist control system and provides an alarm when any local water mist pushbutton is actuated. The machinery control system automatically attempts to complete any unsuccessful actions commanded by the local pushbutton. If the machinery control system is also unsuccessful, the machinery control system operator is notified of the failures in order for the operator to reconfigure the system to complete the water mist operation. The machinery control system will also perform appropriate electrical isolation. Whenever the water mist system is activated from any location, the pump start time is delayed for 15 seconds so the machinery control system can reconfigure the electric plant to electrically isolate the selected space. Personnel can stop or override the reconfiguration of the electric plant and water mist pumps will start after the 15 second delay even if the electric plant reconfiguration is incomplete, changed, or aborted.

APC System

APC fire protection systems are installed in Navy ships and submarines to provide protection for galley deep fryers and their exhaust systems. Aqueous potassium carbonate is specifically formulated to extinguish fire in the reservoirs by combining with the hot cooking-oil surface to form a combustion-resistant soap layer, thereby cutting off the grease from its source of oxygen. There is little or no cooling with APC. The following systems are installed on ships:

1 A system.
2 AA system.
3 B system.
4 Modified B system.

The A system uses a single cylinder assembly loaded with 2-½ quarts or 5 quarts of extinguishing agent depending on cylinder size installed to supply four appliance nozzles. The AA system uses a dual cylinder assembly loaded with 5 quarts of extinguishing agent to supply two sets of four appliance nozzles (3/8-inch discharge tubing is used to supply the nozzles). The A and AA systems are designed for use with appliance nozzles only. The B system uses a single 6-gallon cylinder assembly loaded with 4 gallons of extinguishing agent to supply two appliances, two plenum range hoods, and two duct nozzles (5/8-inch discharge tubing is used to supply the nozzles). The modified B system uses a 6-gallon storage cylinder containing 4 quarts of extinguishing agent. The BB system uses a dual cylinder assembly. Each cylinder is loaded with 4 gallons of extinguishing agent to supply appliance, plenum range hood, and duct nozzles (7/8-inch discharge tubing is used to supply the nozzles). The A system was originally provided with a small 2- ½ quart cylinder. The 2-½ quart cylinder went out of production by 2004 and was replaced with a larger 5-quart cylinder. The APC system is being phase out because deep fryers are being removed or not being installed on newer ships.

Components

Each APC system includes one or two cylinders filled with a solution of potassium carbonate in water pressurized with compressed nitrogen (N_2). Discharge piping from the cylinder(s) leads to one or more nozzles which spray the solution into the cooking oil reservoirs, along the galley hood plenum, or up into the galley hood exhaust duct. A spring-tensioned cable keeps the system inactive. When this tension is released, the system is activated and N_2 is released from a pressurized cartridge. This action opens the lever control heads, releasing the aqueous potassium carbonate. The cylinder assembly plays a major part in the APC system. The cylinder assembly is composed of the following:

1 Cylinders - the cylinder is used to store the system fire extinguishing agent, which is a solution of potassium carbonate and water. The cylinder is charged to 175 psi with nitrogen gas, installed in an upright position, and located near the deep fryer so that it is accessible to the operator. The modified B system cylinder is charged to 100 psi.
2 Cylinder Valve - The cylinder valve connects the cylinder assembly to the discharge system. When the valve is opened the cylinder will discharge the extinguishing agent. A pressure gage is mounted to the cylinder valve so that the cylinder pressure is easily read.
3 Lever Control Head - The automatic or manual discharge of the cylinder assembly is controlled by the lever control head. One-fourth inch copper tubing from the pressure release control box is connected to the lever control head used for automatic operation. The lever control head's lever handle can also be used to manually discharge the cylinder assembly once the release pin is removed.
4 Pressure Switch - Each cylinder assembly is fitted with an appliance pressure switch to interrupt power to the deep fryer when the cylinder is discharged or the pressure charge leaks off. The cylinder assembly pressure switch actuates local alarms when cylinder pressure drops. The other pressure switch is installed in the APC distribution line to actuate remote alarms, secure ventilation, and interrupt power to the fryer.

The discharge system is designed to deliver the fire extinguishing agent to the nozzles. The basic design includes 5/8 or 3/8-inch stainless steel (OD CRES) tubing and a vent plug. The vent plug is installed in a tee positioned in the discharge line and gets rid of any static pressure buildup. The nozzles are used to distribute the extinguishing agent onto the fire. There are three nozzle types. The appliance nozzle discharges onto the protected appliance. The plenum range hood nozzle is centrally located in the plenum with its discharge port directed along the axis of the plenum. Each nozzle can protect up to 10 feet along the length of the plenum. The duct nozzle is located inside of the exhaust duct with its discharge port directed along the axis of the duct. Each nozzle is fitted with a strainer to remove particles that could clog a discharge port. Each nozzle discharge port is also protected by a foil sealing disk to prevent clogging of the port by grease or particles from the fryers.

Operation

Operation of the APC fire-extinguishing system is normally fully automatic. Manual backup modes of operation are provided at the cylinder assembly, pressure release control box, and the remote manual control box. The remote manual control box houses the anchor for the cable release system. The release cable end is anchored to a U-bracket by a release pin. A wire and tamper seal secures the release pin. The release cable system is made up of a wire that runs through sections of conduits and pulleys. The release cable sections are lengths of 1/16-inch cable threaded through sections of conduits. Any change in cable direction is accomplished through the use of pulleys.

The pressure release cartridge consists of a cartridge, operating release lever, valve, and pressure gage. The cartridge is charged to 300 psi with nitrogen. The operating release lever is connected to the release cable turnbuckle and the extension spring. The lever operates the piston valve which is screwed into the cartridge and which controls the flow of nitrogen to the cylinder assembly through 1/4-inch copper tubing. A pressure gage attached to the cartridge provides a means to inspect the pressure.

The control box is the mounting platform for the pressure release cartridge and consists of a box with a hinged front panel containing an extension spring, release pin, and turnbuckle. The extension spring is attached to the operating release lever and is anchored to the wall of the box. One end of the turnbuckle is pinned to the operating release lever by a release pin which is secured with a wire and lead seal. The other end of the turnbuckle is connected to the release cable by a large screw eye.

Automatic Operation

The pressure release control box activates the cylinder assembly during automatic operation. The system can also be discharged manually from the pressure release control box. It includes the pressure release cartridge and the control box. Excessive heat on the fusible links melts the link and releases the cable tension. The fusible links are designed to melt at 360°F and when a link melts, it releases the tension of the cable and allows the pressure release control box to activate the system. The extension spring in the pressure-control box pulls the lever, which activates the pressure release cartridge. N_2 gas from the pressure-release cartridge activates the lever control head(s), causing the cylinder(s) to discharge. The appliance detectors are arranged in a straight line. One appliance detector is provided for each fryer and is mounted to the hood directly over the protected fryer. A duct detector, if fitted, is located in the exhaust duct. The duct detector is arranged in scissors fashion in order to locate the fusible link out into the exhaust system. The fusible links supplied with this system take one to three minutes to melt under optimum conditions.

Manual Operation

The APC system has three manual modes of operation:

1 At the cylinder assembly, remove the release pin in the lever-control head completely, and operate the lever. This discharges the cylinder directly.

2 At the pressure release control box, open the box and remove the release pin completely. This disconnects the release cable and

allows the extension spring to activate the system as described under automatic operation.

3 At the remote manual-control box, remove the release pin completely. This disconnects the anchored end of the release cable, releases the tension, and allows the extension spring to activate the system as described under automatic operation.

Miscellaneous Seawater Sprinkler Systems

Miscellaneous seawater sprinkler systems are normally installed in spaces where the quantity and combustibility of materials present is high enough that, should a fire occur in these materials, hose line attack would not succeed in preventing compartment burnout and major damage. These systems may be installed in certain spaces including incinerator rooms, dry stores cargo holds, storerooms, cargo holding areas, living spaces, aviation tire storerooms, flammable gas cylinder storerooms, lubricating oil store-rooms, trash compactor rooms, paint spray rooms and paint spray booths, solid and plastic waste reprocessing rooms, rubber and plastic shops, and other high fire risk spaces. Except for paint spray rooms, paint spray booths, lube oil storerooms and flammable gas cylinder storerooms, ordinary combustibles constitute the major fire risk in spaces with miscellaneous sprinklers. A sprinkler system is also installed outside and around vital spaces of some ships to prevent fire spread into the vital space. Ship designs with a large population of miscellaneous wet type sprinkling systems include guided-missile destroyer (DDG) 51 and LPD 17. Ships with a large population of dry type miscellaneous sprinkling systems include dock landing ship (LSD) 49 and CVNs with tire storage compartments.

Controls

Sprinkling control valves are manually operated local to the compartment protected and may also be provided with remote seawater hydraulic control on the damage control deck. They are typically provided with a water switch (dry type system) or flow switch (wet type system), which sends an alarm signal to a remote damage control panel when the system operates. A local audible alarm is not provided.

Manual Dry Type Miscellaneous Sprinkling System

The manual dry type sprinkler system does not contain seawater downstream of the normally-closed sprinkling control valve when the system is inactive. Dry-type systems are actuated by ship's force manually opening the normally-closed sprinkler control valve, allowing seawater to flow to the sprinkler piping. Sprinkler heads may be either open head or closed head (Figure 4-7). When closed head sprinklers are used, only those heads that are opened (fused) by heat from a fire will flow water. When open heads are used, all heads flow water when the sprinkler control valve is opened. More sprinkler heads are expected to open with a dry type system with closed heads than with a wet type system with closed heads, because the delay for human actuation of the system, so wet type systems are now preferred. Dry type systems represent older designs, no longer specified, that were commonly found in flammable gas cylinder storerooms and aviation tire storerooms.

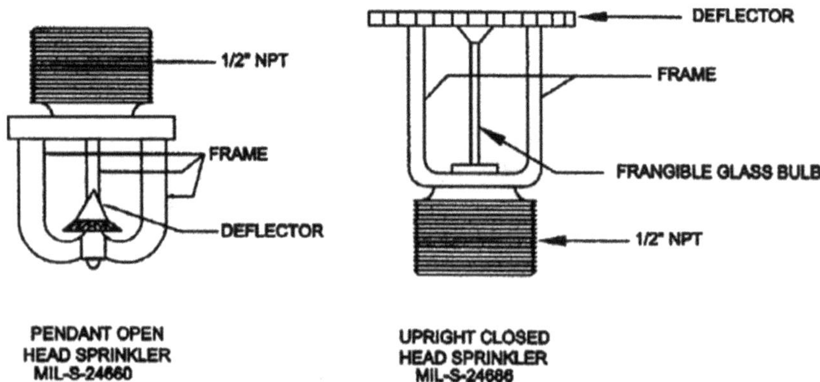

Figure 4-7. Open and Closed Sprinkler Heads

Automatic Wet Type Miscellaneous Sprinkler Systems

The automatic wet type miscellaneous sprinkler systems contain seawater throughout the piping down to the sprinkler heads. The sprinklers are the closed head type and will actuate automatically when exposed to heat, typically at 175°F (79°C). The sprinkling control valve of a wet type system must remain open at all times except when performing maintenance or system tests. These systems automatically detect heat, initiate an alarm in DCC and begin suppression. Typically, in the early stages of a fire, only a few sprinkler heads open and discharge water. With sprinklers, there are seldom problems of access to the seat of the fire or of interference with visibility for firefighting due to smoke. The downward force of water discharged from sprinklers will lower the smoke density in a room where a fire is burning and will also serve to cool the smoke making it possible for persons to remain in the area much longer than they could if the room were without sprinklers. Closed head wet type sprinklers are expected to limit fire damage and operate before flashover. These systems tend to wet down adjacent combustibles preventing fire involvement while maintaining a cool overhead thus limiting the number of sprinklers that open.

This chapter introduced the design and function of the major fire protection systems installed in Navy ships. As previous chapters have mentioned, training is vital to the ship's survivability. Extensive training on fire protection systems is held for all shipboard personnel. All crewmembers shall be able to operate the firemain systems, AFFF systems, carbon dioxide (CO2) systems, halon systems, APC system, water mist and various sprinkler systems. It is very important to utilize any of these systems to immediately gain control or extinguish a fire that can be fatal to the ship. The next chapter will cover firefighting systems and equipment used on U.S. Navy submarines.

CHAPTER 5

Submarine Fire Protection Engineering

Submarines are designed to absorb and recover from all types of damage. A well-structured and knowledgeable firefighting and emergency response organization will depend on the maximum efficiency from the survivability features designed into the submarine. The control of damage and causalities on a submarine is always an all hands evolution. The submariner navy does not have a specialized fire protection engineer such as surface ship's damage controlman and the aviation community's aviation boatswain's mate. All personnel that are attached to submarines must become thoroughly familiar with fire protection engineering. The submariner firefighting and emergency response organization have very similar positions and functions as the firefighting and emergency response organization on ships. The submariner organization must be very flexible to combat various fires and causalities under various circumstances. This chapter will discuss the roles and responsibilities of the firefighting and emergency response organization and the fire protection systems and equipment used for survivability of submarines.

Firefighting and Emergency Response Team

Because of the unique characteristic of a submarine, it can immediately become a blazing inferno in the matter of minutes and rapid response is essential to control the fire. All personnel play a major role in the submarine's survival. As previously stated, the firefighting and emergency response teams on submarines are very similar to the organizations found on ships. The firefighting and emergency response organization is made up of the following positions and teams:

1 Casualty coordinator.
2 Damage control assistant (DCA).
3 Man in charge at the scene.
4 Firefighting teams.
5 Rapid response team.

Casualty Coordinator

During any fires or emergency, the casualty coordinator will report to DCC to direct the overall coordination of submarine's firefighting and emergency response actions. The submarine's navigation officer or engineering officer will assume responsibilities of the casualty coordinator. The engineer officer will assume casualty coordinator for emergencies outside engineering spaces. The navigation officer will assume casualty coordinator for emergencies in the engineering spaces which allows the engineer officer to respond to emergencies that affect the engineering plant.

Damage Control Assistant (DCA)

The DCA, under the supervision of the engineer officer is responsible for maintaining the ship's fire bill, maintenance of firefighting and damage control equipment, and the training of the firefighting and emergency response team. The role of the DCA on submarines slightly varies from the DCA role on the surface ships. The DCA on surface ships will direct all firefighting and damage control actions while on submarines, the DCA will assume the role as assistant casualty coordinator.

Man at Charge at the Scene

The man in charge at the Scene also referred to as the OSL is in charge of firefighting operations at the scene of a fire. The senior man in the affected compartment acts as the OSL until relieved. The designated OSL is usually the XO when the submarine is underway or the ship's duty officer when the submarine is in-port. The OSL has similar responsibilities as surface ships OSL but specifically will conduct the following actions during fires and emergencies:

1 Directs the attack against fires.
2 Directs the electrical isolation of equipment, as required.
3 Establish communications with the casualty coordinator
4 Follow coordinated instructions from the casualty coordinator.
5 Determines what resources are needed for the firefighting team.
6 Determines the number of personnel needed for the hose teams.

Firefighting Teams

Submarines have a minimum of four firefighting teams during underway conditions and one firefighting team per duty section for inport periods. The number of personnel on each firefighting team will vary depending on the assigned number of personnel available on each submarine. The firefighting team will consist of a nozzleman and two hoseman. The use of a team leader will be determined by the OSL. The OSL will assign team leader responsibilities to the nozzleman or to an additional personnel individual when necessary. When a separate individual is assigned as team leader, they will operate the NFTI during firefighting operations. A separate individual as team leader is desirable to direct the rotation of firefighting team personnel. The team leader shall direct the hose team to do the following:

1 Directs hose team to charge the fire hose.
2 Directs the nozzleman in the selection of fire nozzle spray patterns.
3 Directs the hose team members in the movement of the hose.
4 Directs the hose team members to rotate and relieve the nozzleman.

Rapid Response Fire Team

Successful fire extinguishment depends on the immediate attack on the fire. The rapid response fire team fills the gap between the time the fire is discovered by a crew member and the time the firefighting team relieves the rapid response team. The rapid response fire team must be relieved as soon as possible by the firefighting team. Each submarine will have a designated rapid response fire team for underway conditions. As a minimum, the following functions will be performed:

1 Immediately proceed to the fire and combat the fire with portable extinguishers, fire hoses or AFFF hose reel (when available).
2 Establish communications.
3 Rig fire hoses.

Submarine Fire Fighting Systems and Equipment

Submarines have some unique characteristics that affect fires and the spread of smoke and heat from a fire. The primary characteristics unique to a submarine that affect fire growth are the closed environment of a submarine and the oxygen concentration. Fires tend to become intense quickly in the closed environment of a submarine. The insulated hull reduces the transfer of heat from the submarine to the environment. This increases the amount of heat from the fire contained inside the submarine.

Also, the pressure increase caused by the fire may contribute to the rapid growth of the fire. Fire protection systems and equipment used on submarines provides personnel with a rapid response to extinguish a fire before it gets out of control. The fire protection equipment and protective gear used on ships are also used on submarines. The fire protection systems and equipment particularly used on submarines are as follows:

1 Trim system (firemain).
2 AFFF distribution system.
3 Portable AFFF injection unit (PAIU).
4 Seawater flooding and sprinkler systems.
5 Freshwater hose reel system.
6 APC system.
7 Missile gas system.
8 Portable electric submersible pumps.
9 Emergency air breathing (EAB) system.
10 Ventilating systems.

Trim System (Firemain)

As mention before, the sea provides an inexhaustible amount of water and the firemain system is filled with seawater pumped from the sea. The firemain system onboard submarines are known as the trim system. The trim system consists of pumps, piping, tanks, and valves through which seawater is supplied to fire hose stations, AFFF systems, and sprinkling systems that provide water for flooding the pyrotechnics, small arms ammunition, and chlorate candle stowage lockers. The trim systems are basic firefighting system used on submarines except for the following systems:

1 SSN-21 (nuclear-powered general purpose attack submarine) class firemain system.
2 SSN-774 class firemain system.

SSN-21 Class Firemain System

SSN-21 class firemain system consists of one firemain system installed in forward compartment and another in the after compartment of the submarine. The firemain is supplied with sea water from the pressurized trim tanks and distributes sea water to services as required. The forward firemain is installed between the forward trim tank and the trim discharge header near the forward reactor compartment bulkhead. The after firemain is installed between the aft trim tank and the trim discharge header near the after reactor compartment bulkhead. Valves are installed in the trim discharge header to permit the forward trim tank to pressurize the aft firemain with the aft trim discharge header isolated, and vice versa.

The SSN-21 class firemain system is designed so fire plugs can maintain a minimum operating pressure of 75 psi when two Navy standard fire hoses (1-¾ inch or 1-½ inch) and the largest sprinkling system is operating simultaneously. Fire hose stations are located such that all parts of the forward compartment and engine room can be reached from at least two stations with 50 feet of hose. Additional hose lengths are stowed in the engine room on hose racks to allow two hose coverage to any part of the reactor compartment. Reactor compartment coverage is initiated by connecting any stowed hoses to existing 50-foot hose lines and fireplugs located in the engine room. All fire hose stations are also provided with AFFF capability.

SSN-774 Class Firemain System

The SSN-774 Class system consists of two firemains located forward and aft sections of the submarine. AFFF solution can be injected into this firemain system and distribute AFFF to fire hose stations, hose reels, and the diesel overhead sprinkler system. The AFFF is supplied by the portable AFFF injection unit (PAIU) which will be discussed later in this chapter. This system also provide seawater flooding to the aft chlorate candle locker in the engine room, the small arms ammunition (AMMO), and pyrotechnics (PYRO) lockers located in the forward compartment. This system also provides sprinkling to lockout trunk (LOT).

AFFF Distribution System

The AFFF distribution systems installed on the SSN-21 class submarines is designed to supply AFFF to sprinklers that protect the diesel generator space, the lube oil bay bilge, the areas outboard of the fan room, and the area around hydraulic plants in the engine room. The AFFF system also provides AFFF to all 1- ½ inch seawater fireplugs. AFFF is generated from two AFFF distribution systems installed forward and aft. The forward AFFF system supplies the forward compartment fireplugs and sprinklers and aft AFFF system supplies the engine room fireplugs and sprinklers. The AFFF distribution system consists of an AFFF concentrate tank, AFFF concentrate supply piping, proportioning equipment, and AFFF piping that supplies AFFF solution to sprinklers, check valves, isolation valves, and controls. The concentrate tank is constantly pressurized from the submarine's service air system at a sufficient pressure that will inject AFFF into seawater supplied from the firemain. The Type 3 AFFF concentrate is injected through constant flow control valves into the firemain branch lines supplying fireplugs and sprinklers to create a 3 percent AFFF to 97 percent seawater solution.

AFFF capability is provided to all seawater fireplugs. AFFF concentrate is injected into the firemain branch line of each fireplug from a PAIU through a constant flow control valve (Figure 5-1). Each AFFF concentrate supply line has an isolation valve, a check valve, fireplug AFFF cutout valve, and constant flow control valve upstream of the AFFF injection point. Each firemain branch line has a check valve and a fireplug cutout valve upstream of the AFFF injection point to eliminate AFFF backflow into the firemain. Each AFFF equipped fireplug is capable of supplying seawater or AFFF solution regardless of whether other AFFF stations have been activated. Each AFFF distribution system is sized to support simultaneous operation of the largest sprinkling load in the distribution system's respective compartment and two Navy standard fire hoses (1-¾ inch or 1-½ inch) for 10 minutes. The AFFF solutions produced through the constant flow control valves has an AFFF concentration level of at least 3 percent and no greater than 4 percent.

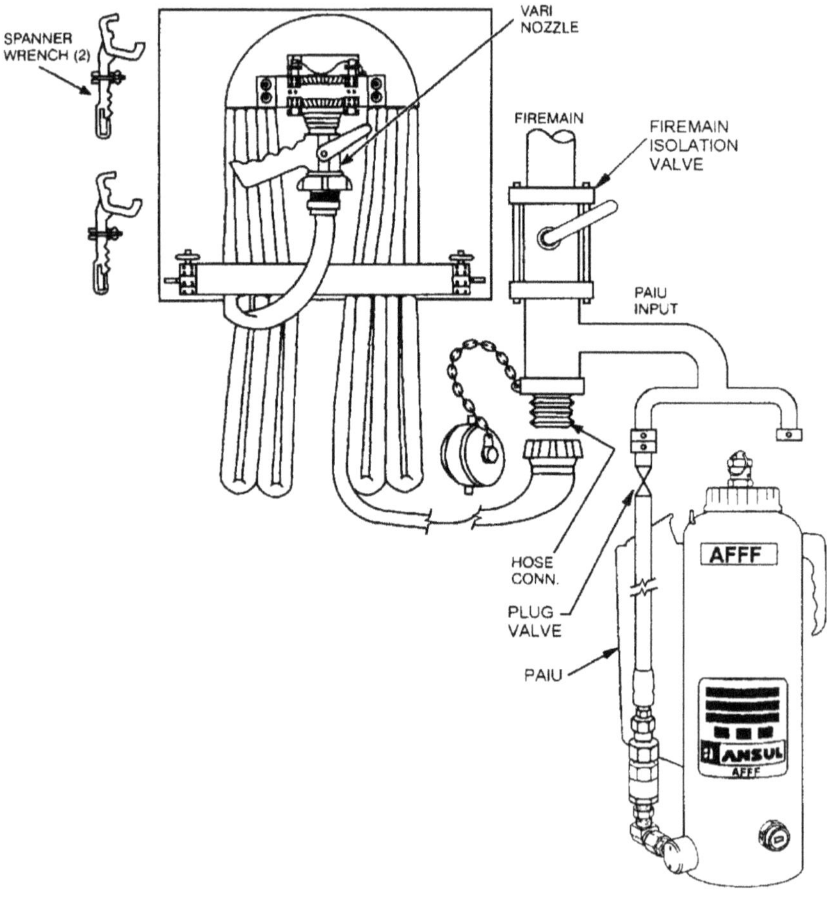

Figure 5-1. Fire Hose Station with Portable AFFF Injection Unit (PAIU)

Service Outlet Operation

To operate the fireplug hose with seawater only requires manual operation of the fireplug cutout valve and hose nozzle. To operate the fireplug hose with AFFF solution requires manual operation of the fireplug AFFF cutout valve, fireplug cutout valve, and the hose nozzle.

AFFF Sprinkler Systems Application

The AFFF sprinkling systems installed on submarines to provide AFFF bilge coverage whenever there is a fuel spill. The submarine AFFF sprinkler systems are located in the following areas:

1. Diesel generator space - Sprinklers are installed under the deck plates to protect the bilge region. Additionally, a minimum of two directional, fan-spray pattern sprinklers are installed on each side of the diesel engine, angled so that AFFF covers the vertical surfaces of the engine but does not overspray onto the generator end of the diesel and adjacent equipment. The sprinklers protecting the diesel are supplied from the same sprinkler control valves protecting the bilge region. A manual shutoff valve is located outside the generator compartment to secure the directional sprinklers and allow operation of the bilge sprinklers only.
2. Lube oil bay bilge - Sprinklers are installed under the deck plates to protect the bilge region.
3. Areas around the hydraulic plants - Sprinklers are installed around each hydraulic plant in the engine room to protect the tank top and any surrounding bilge region.
4. Areas outboard the fan room - Due to access limitations, sprinklers are installed outboard the fan room and directed towards the frame bays to protect the area from fires coming from the machinery space below. AFFF concentrate is injected into the firemain branch line of each sprinkling system through an isolation valve, a remotely operated control valve, and a constant flow control valve upstream of the AFFF injection point. Each firemain branch line supplying the sprinkler system has an isolation valve and a remotely operated control valve upstream of the AFFF injection point.

AFFF Sprinkler System Operation

The sprinkler control systems are activated by manual seawater hydraulic operated control valves. Each AFFF sprinkler control system operates the sea water control valve and the AFFF concentrate control valve for the AFFF sprinkler system. A local manual control valve is located outside the engine room and inside the engine room in the proximity to the protected areas. A remote manual control valve is located at the AFFF system control panels located in DCC. The fan room sprinklers are supplied from a Navy standard fire hoses when connected by personnel. The hose connections are located inside the fan room.

The AFFF system control panel is located in DCC. The AFFF system control panel consists of the following:

1 Remote manual control valves for operating the respective sprinkler systems.
2 Audible signals to indicate operation of the respective sprinkler system remotely operated control valves and the low level alarm.
3 Visual signals to show "open" or "shut" position of the respective sprinkler system remotely operated control valves and the tank low level alarm.

Portable AFFF Injection Unit (PAIU)

The PAIU consists of the cylinder assembly and hose assembly (Figure 5-2). The 3-gallon cylinder has a fill cap with relief valve, carrying handle, contents indicator, discharge fitting, and gas pressure cartridge. The CO_2 gas pressure cartridge mounted to the extinguisher shell charges the cylinder through a pressure regulating valve. The discharge hose assembly consists of a check valve, flow control valve, hose, plug valve, and quick disconnect fitting. The hose assembly is ten feet in length. The PAIU is filled with 2.5 gallons of 1 percent AFFF concentrate and will discharge in approximately 2 to 2.5 minutes when connected to a fire hose station. The PAIU will provide an extended discharge time of approximately 6 to 8 minutes when connected to a hose reel or the diesel generator sprinkler system due to an additional flow control device in the station piping line.

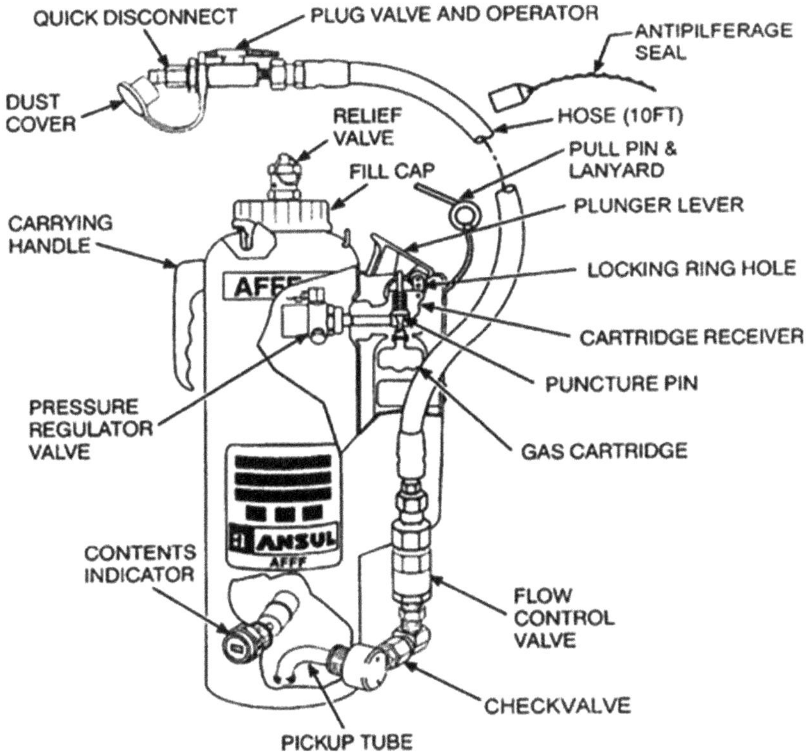

Figure 5-2. Portable AFFF Injection Unit

The portable AFFF injection unit (PAIU) is designed to store and deliver 1-percent AFFF concentrate injected into the AFFF and trim system (Figure 5-1). The PAIU injects AFFF concentrate through quick disconnect fittings at each fire station and fire hose reel. To establish a continuous operation of AFFF, two PAIUs can be connected to the fire station. As the first PAIU is completely discharged, a second PAIU can be activated to provide a continuous flow of AFFF. Additional PAIUs can be placed at the station and connected to replace the discharged cylinders as needed. The PAIU is manufactured by Ansul Incorporated and uses

1-percent ANSULITE concentrate that is currently meets the fire performance requirements of Underwriters Laboratories Standard (UL-162, UL Standard for Safety Foam Equipment and Liquid Concentrates). The PAIU is not a stand-alone portable fire extinguisher and must be used with a hose station, hose reel, or the diesel sprinkler system to provide properly mixed AFFF solution.

Seawater Flooding and Sprinkler Systems

If a fire threatens the pyrotechnic and small arms ammunition lockers, these lockers can be flooded by seawater by the means of a single locked box flood valve which receives seawater from the trim system on submarines other than the SSN-21 and SSN-774 classes. The valve and its operating wrench are located inside a breakable, glass box located within the compartment containing these lockers or remotely from a valve located on the lockers. On the SSN-21 and SSN-774 classes, the seawater supply to the pyrotechnic and small arms ammunition lockers has piping from the trim system with a seawater valve that controls the seawater flow.

An overhead sprinkling system is provided inside the oxygen chlorate candle stowage lockers on the SSBN-726 (nuclear-powered ballistic missile submarine), SSN-21, and SSN-774 class submarines. The lockers are equipped with sprinklers that will cool the candles to prevent their ignition in event of a fire in or near the locker. The chlorate candle lockers receive water from the trim system through locked box flood valves similar to the system for flooding the pyrotechnic and small arms ammunition lockers on the SSBN-726 class. On the SSN-21 class, the seawater supply piping is controlled by seawater valve that allows seawater to flow from the trim system. On SSN-774 class submarines flooding of the forward chlorate candle storage is performed with the nearest available fire hose or hose reel. The aft chlorate candle locker is flooded via a dry standpipe fitting supplied from fire hose connected from fire hose station on engine room lower level.

For submarines without locker sprinkling, application of water fog will extinguish a candle fire. Water fog may be used to cool a locker when exposed to a nearby fire. Submarine chlorate candle lockers on future submarine classes no longer require fixed sprinkler protection unless a nearby fire hazard is identified that would prevent access to the locker to apply firefighting water. The reason sprinkling applications are no longer used for new submarines is that past fire incidents involving candles often occurred in the furnace not the storage locker.

Freshwater Hose Reel System

Freshwater hose reels are installed in SSN-21 class submarines to provide a rapid response firefighting capability for Class A fires. Hose reels are located such that all parts of the forward compartment and engine room can be reached with at least one hose with minimal travel up or down ladders. Two hose reels have 75-foot long, 1-inch diameter non-collapsible, rubber hose. The 1-inch diameter hose has a ¾-inch nozzle. All portions of the reactor compartment are reached with freshwater for firefighting from a connection in the vicinity of the reactor compartment access door. The freshwater hose reel system receives water from the submarine's potable water system at a minimum pressure of 50 psi. The freshwater hose reel nozzle will produce 15 gpm at 25 psi.

APC System

As previously stated in chapter 4, the APC fire extinguishing system is installed in submarines to provide protection for galley deep fat and doughnut fryers and their exhaust systems. The APC systems on ships and submarines have similar components. There are two basic APC fire extinguishing systems installed on submarines. The systems are designated as B and modified B. The B system that is installed on the SSN-688, SSBN-726, SSN-21, and SSN-774 classes uses a single 6-gallon cylinder assembly loaded with 4 gallons of extinguishing agent to supply appliance, plenum hood, and duct nozzles. The modified B system installed on the SSN-637 class uses a 6-gallon storage cylinder containing 4 quarts of extinguishing agent.

The remote activation for the APC system may slightly vary on each submarine class. Specifically the SSBN-726, SSN-21, and SSN-774 class submarines, the system can be activated remotely at the centralized galley fire control station. The SSBN-726 class submarines' APC can be activated by depressing the remote actuator detent button and pushing in the actuator. The SSBN-726 class submarines' APC can be activate by depressing the remote actuator detent button and pushing in the actuator. The SSN-21 and SSN-774 class submarines' APC can be activated remotely by pulling the release pin at the remote manual control box.

Missile Gas System

The missile gas system is fire suppressing system that is installed in submarine's missile launching tube. The system uses nitrogen to pressurize the launching tube during missile launches or fail launches. Nitrogen will not support combustion of the launch gases and can act as an extinguishing agent by inerting the atmosphere to smother or prevent a fire in the missile tube.

Portable Electric Submersible Pumps

The portable electric submersible pump (ESP) is a centrifugal pump driven by a water-jacketed constant speed ac electric motor. It is designed to operate as single-phase or three-phase at 120, 240, or 440 volts. This design is rated to deliver 140 gpm against a maximum head of 70 feet and 180 gpm at a 50-foot static head. The output is variable and will increase with a decrease in head pressure. This portable electric submersible pumps are supplied to ships as a means for removing water from flooded compartments. It is used on submarines primarily for drainage of isolated bilge pockets which cannot otherwise be drained under extreme angles of trim. The pump can be used to fight fires in emergency situations where there is a need for more firefighting hose lines. The pump will produce a low discharge pressure that will not operate the vari-nozzle in a fog position, but can produce enough pressure for a straight stream firefighting pattern with a 30 to 40-foot reach.

Emergency Air Breathing (EAB) Systems

The emergency air breathing (EAB) system is designed to supply oxygen to personnel conducting emergency operations in a toxic atmosphere such as smoke, vapors, fumes, or gases. The EAB system consists of installed air breathing manifolds and air-line face masks. The EAB air manifolds provides clean, filtered air at 100 psi to personnel engaged in firefighting and emergency situations for an extended stay in a compartment containing unbreathable air and poor visibility. The EAB stations are clearly labeled with posted operating instructions. The EAB mask plugs into the EAB stations for emergency breathing air supply. The air-line mask is to be used by rapid responders on the scene for firefighting until relieved by firefighting team members wearing SCBAs.

When EAB is required because of environmental contamination, the face mask with its attached demand regulator is donned and the hose is connected to the manifold. The demand regulator reduces supplied air at 100 psi to breathable levels in the mask as the wearer inhales. If necessary, movement between stations is possible by disconnecting from one manifold and connecting to another manifold or by connecting to the buddy connection on another wearer's demand regular. The dual action filters provide filtered air to the EAB manifolds.

Ventilating Systems

The ventilation systems consist of vent ducting and fan units that circulate a clean air conditioned atmosphere throughout the submarine. The recirculate mode is a closed loop operation which provides the air conditioned atmosphere in the submarine while keeping the submarine's atmosphere completely isolated from the outside. The recirculate mode is the normal submerged operating mode of the ventilation system. The emergency ventilate mode is the ventilation means used for evacuation of the atmosphere in specific compartment or compartments. This mode is used whenever the submarine atmospheric contain contaminants such as smoke, toxic gas, or other objectionable gases. The emergency mode is an open loop mode that supplies fresh air that is brought into the submarine through the snorkel induction system while air from the contaminated space is exhausted overboard using either the diesel generator or blower. Surface ventilate is an open loop operating mode that is normally used when the submarine is surfaced for an extended period in calm seas and when in port.

When a submarine has a fire emergency, immediate action must be taken to isolate unaffected compartments. The ventilation system must be secured to prevent smoke from completely filling the entire ship in a short time period. In the event of a Class B fire, the fire should be completely extinguished, a reflash watch set, and overhauled completed before preparations for emergency ventilation of the affected compartment are commenced. During class A or C fires, emergency ventilation may commence after the fire is reported out. If heavy smoke and heat is presence in the affected compartment it may be necessary to ventilate the heat and smoke overboard to assist the firefighting team in combatting the fire. The ship's installed ventilation system should normally not be used in the burning compartment as the intense heat may cause secondary fires along the ventilation path in non-affected compartments. The decision to ventilate the burning compartment before the fire is out will be made by the commanding officer.

This chapter introduced the fire protection engineering organization and the fire suppression systems installed in U.S. Navy submarines. As previous chapters have mentioned, training is the key to the ship's survivability. Training is very vital to submarines because of their special composition. Extensive training on fire protection systems and firefighting tactics is held for all submarine personnel. All submariners shall be able to operate the trim systems, AFFF distributive system, APC system, various sprinkler systems, and portable firefighting equipment. It is very important to utilize any of these systems to immediately gain control or extinguish a fire that can be fatal to a submarine. Rapid response fire team is a critical element in the firefighting and emergency response organization. The rapid response team must be able to control the fire until they are relieved by the firefighting team. The next chapter will cover basic firefighting tactics utilizing firefighting systems and equipment.

CHAPTER 6

Shipboard Firefighting Tactics

As the crewmembers become more proficient in firefighting, combat evolutions, and dealing with engineering plant casualties, they develop the ability to handle more than one single casualty at a time. Their training prepares them for cascading or multiple casualties. The opportunity to conduct mass conflagration training is very crucial to ships. This worst-case scenario helped the *USS Cole* during their attack while docked in Yemen. Also this training helped numerous ships survived multiple missile hits, mine explosions, collisions, and flooding. Fire protection engineer and the crewmembers' ability to think clearly in the face of multiple casualties may someday save the ship. Clear thinking has led to procedures such as using a firehose to jumper around a rupture pipe in the firemain system or using the P-100 pump to replace fire pumps when power is loss to the ships electrical system. Both methods are used to supply firefighting water when the firemain system is damaged or not working. There are many training scenarios to use to prepare firefighting and emergency response team members for the worse case situation.

Ships like the *USS Forrestal* help revolutionized firefighting methods and equipment. Because of the sacrifice others made, future missions can be carried out and lives can be preserved. Training is the key to survival and the fire protection engineer plays a pivotal role with ensuring the crewmembers learn what to do when any emergency occurs. There is never a perfect situation during a fire but the perfect situation for a fire is prevention and training. A shipboard firefighter will most likely encounter different types of fires aboard their ship. Although fires have several things in common, each fire has its own unique characteristics such type of material burning, the way the fire can be isolated, or the space or compartment that has the fire. There are many factors to consider when deciding what tactics to utilize when attacking a fire. It is critical for the fire protection engineers to train the ship's firefighting and emergency response teams to respond to a variety of fire situations. Previous chapters covered the characteristics of fires, fire extinguishing agents, fire protection system and the equipment utilized to fight each type of fire.

Firefighting Strategies

Fires can spread in many different ways as radiant heat from an intense fire may ignite materials in an adjacent compartment, or it may travel through inoperative ventilation ducts to other compartments. Openings between compartments, including cableways, may contribute to the spread of fire. The first sign of the fire spreading is smoke. The investigators and boundarymen play a vital role by ensuring boundaries are established and a continuous investigation is conducted in areas surrounding the primary fire boundaries. The ship's battle organization must ensure that all personnel assigned as investigations and boundarymen are well qualified for their job. They must also report any encounters with any smoke outside the primary fire and smoke boundaries. Then use their breathing apparatus to investigate, if possible. If the fire spreads, then the secondary boundary becomes the primary boundary, and personnel must attack this new threat to the ship.

It is the job of the firefighting and emergency response leaderships to make firefighting decisions that are based on reports received from the OSL, investigators, and boundarymen. A small fire can quickly become a blazing inferno in a very short notice, immediately engulfing an entire compartment or area of the ship or submarine. When the ship is underway, the strategies and methods used in structural firefighting ashore will not work. The crew cannot wait for the fire department to arrive to extinguish fires because they are the fire department.

Fire Properties

There are four classifications of fire and each type of fire has its own distinctive properties and dynamics. As mention in chapter 3, the four classes of fire are the following:

4 Class ALPHA (A).
5 Class BRAVO (B).
6 Class CHARLIE (C).
7 Class DELTA (D).

Class ALPHA (A) Fire

Recalling information for chapter 3, a Class A fire is any fire in which the burning material leaves an ash such as paper, wood, and cloth. These combustible materials are located throughout the ship and submarine. These solid fuels must be heated to their ignition point before they will burn and there must be enough oxygen to support the fire. For a solid fuel to burn, it must be changed into a vapor state by a chemical process known as pyrolysis. The solid fuel decomposes from heat and begins to create a fuel vapor that mixes with oxygen producing a fire.

Removal of any one of the three elements of the fire triangle (heat, oxygen, and fuel) will extinguish a fire. A common method of attacking Class A fires is the application of water. The water cools the fuel below its ignition point, thereby removing heat from the fire triangle which extinguishes the fire. On larger Class A fires, AFFF will be more effective than seawater. In all such fires, other nearby combustibles (including unseen materials on the adjacent and other side of that bulkhead) must either be moved or kept cool to prevent further spread of the fire.

Class BRAVO (B) Fire

A Class B fire presents challenges not confronted in other classes of fires. This is because it can be fueled by any of the flammable liquids stored aboard the ship such as fuels, liquid lubricants, and solvents. Class B fires may be extinguished utilizing halon, HFP, AFFF, water mist, PKP, or a combination of these agents, but the single most important step in fighting this monstrosity is to secure the source of the fuel. One of the characteristics of a flammable liquid is the flashpoint which is the lowest temperature at which the liquid will give off sufficient vapor to form with an ignitable mixture. When mixed with air, this vapor can ignite from an ignition source small as a spark.

Fuels and other liquids stored aboard ship are often pressurized (to pump them to other areas of the ship), or may be stored under pressure to minimize the release of vapors. Leaks in these pressurized fuel systems will tend to spray outward becoming atomize increasing the possibility of coming into contact with an ignition source. As an example, the ignition source could be a heated surface in an engineering compartment or an electrical spark from a faulty electrical component. When there is a flammable liquid spill or leak from a pressurized source, the leak or spill will cover a large area. When this occurs a great amount of vapor will be released. If it ignites the fire will produce a great amount of heat. One of the specifications of flammable liquids is that each liquid has a minimum flashpoint. Anytime a ship is refueled, the fuel must be tested for both quality and flashpoint before it is received from the supplier.

Some flammables require special storage, often in special lockers with temperature detection and sprinkler systems installed. Some of the materials stored in these lockers are paints, welding gases, flammable cleaning solvents, and other materials. An accurate inventory of hazardous materials stored in such lockers should be readily available. Fuels for portable P-100 pumps and small boats may sometimes be stored on the weather decks of the ship in authorized containers. The ship's supply or safety department can provide information about flammable materials (including safety and handling precautions, hazards, and minimum flashpoints). The Safety Data Sheets (SDS) has information on each individual hazardous product carried onboard ship.

Class CHARLIE (C) Fire

A Class C fire is an energized electrical fire and it may be attacked with nonconductive extinguishing agents such as CO_2 or with low-velocity water fog. Special care must be taken to maintain a safe distance from energized equipment. The most common and safest method of dealing with a Class C fire is to secure the electrical power source and treat it as a Class A fire by extinguishing the burning wire insulation. Special care must be taken to avoid contact with energized electrical equipment. The user must remember to round the CO_2 bottles and keep the horn of the portable extinguisher from contacting the energized equipment. If it is necessary to use water fog as an extinguishing agent, a minimum distance of 4 feet must be maintained while applying the water fog repetitively by opening and closing the nozzle in a repeated action until the fire is extinguished. A straight stream of water must never be used on a Class C fire due to possible electrical shock.

Class DELTA (D) Fire

Class D fires are also known as combustible metal fires such as magnesium, phosphorus, sodium, or titanium. Certain types of aircraft wheels are manufactured from these materials as well as various pyrotechnic materials. Although some ships have pyrotechnic ("pyro" for short) magazines below decks, typically most occurrences of Class D fires are located topside where storage is more common. Pyrotechnics often contain their own oxidants and therefore do not depend on atmospheric oxygen for combustion. So removing the oxygen by utilizing agents such as PKP, AFFF or halon 1211 extinguishers will be ineffective.

Class D fires burn with an intense heat of temperatures up to 4,500°F and produces a very bright light that can damage personnel's eye sight. High velocity fog should be used to cover and cool these fires, but if possible, remove the burning material by throwing (jettisoning) it over the side of the ship. To prevent the fire from spreading, the firefighting team should apply large quantities of water at low pressure to cool the surrounding area. Class D fires give off extreme amounts of heat and can produce explosions. Because of this danger, the firefighting team must maintain a safe distance from the source of the fire while applying the water fog. During a Class D fire certain chemical reactions will occur as the water is applied to cool the surrounding area. This water reacts with the burning metal and forms hydrogen gas which will either burn or explode, depending on the intensity of the fire and the amount of burning material. In any case, the firefighting team must maintain a safe distance and shelter from any potential explosions cause by the fire.

Dynamics of Fire

It cannot be overemphasized that there is an extensive assortment of combustible materials aboard any ship and submarine which can be quickly ignited to cause a major catastrophe. As stated in a previous paragraph, in order for a solid fuel to burn, it must be changed into a vapor state through a chemical action known as pyrolysis which is defined as a chemical decomposition due to the application of heat. A fire is produced once the fuel vapor is mixed with oxygen and heated to the specified temperature. Solid fuels will burn at different rates depending upon its size and configuration. For example, a pile of wood chips or wadded paper will burn faster than an equal amount of solid wood or a case of paper. This is factual because there is a larger surface area exposed to the heat; therefore, vaporization occurs faster. Because more vapors are available for ignition, the fire will burn more intensely and the fuel is consumed much faster.

A liquid fuel will release vapor as much as a solid fuel does, but at a higher rate and over a larger temperature range. Because liquids have loosely compacted molecules, heat increases their rate of vapor release. This is the reason why liquid fuels give off heat quicker and they produce about 2 ½ times more heat than wood. For example, if a flammable liquid is spilled or become atomized by spraying out under pressure, a very large surface area will be affected. The flammable liquid will give off much more vapor. Because of these characteristics, this is why Class B fires burn so violently.

As mentioned in a previous paragraph, the lowest temperature at which a liquid gives off sufficient vapor to form an ignitable mixture is known as the flashpoint for that liquid. An ignitable mixture is a mixture of vapor and air that is capable of being ignited by an ignition source. For example, gasoline has a flashpoint of -45°F (-43°C) making it a constant hazard because it produces flammable vapor at normal atmospheric temperatures. Just like gasoline, all the other shipboard fuels have specified minimum flashpoints.

To ignite a flammable gas or vapor, it must be mixed with the air in the proper ratio. The lowest percentage of gas that will make an ignitable mixture is called the lower explosive limit (LEL). If there is less vapor or gas than this percentage, then the mixture is too lean to burn. Also there is also an upper explosive limit (UEL) in which the mixture is too rich to burn. The range between the lower and upper explosive limits is called the explosive range. Here is a quick example. A mixture of 10 percent gasoline vapor and 90 percent air will not ignite, because the mixture is too rich (above the UEL).

Fire Growth

A quick response is needed to prevent a fire from engulfing a large area and becoming uncontrollable. When a fire ignites in shipboard compartment, a fire goes through four distinct stages of life. These stages are known as the following:

1 Growth Stage.
2 Flashover Stage.
3 Fully-Developed Fire Stage.
4 Decay Stage.

Growth Stage

During the growth stage of a fire, the average compartment temperature is low and the fire is localized near its origin. The rising heat from the fire and the smoke forms a hot upper level in the affected compartment. When this continues, a flame front of burning gases is formed across the overhead of the compartment. This process is known as rollover which takes place when unburned combustible gases from the fire mix with fresh air in the overhead and begin burning at some distance from the fire. It is important that the firefighting team check the overhead or ceiling for rollover as soon as they get access to the compartment. Rollover differs from flashover in that only gases are burning in the space not all the contents of the space.

Flashover Stage

The flashover stage is the period of transition from the growth stage to the fully-developed fire stage. It occurs in a short period of time and may be considered an event, much as ignition is an event in a fire. It normally occurs at the time the temperature of the upper smoke layer reaches 1100°F (600°C). The most obvious characteristic of flashover is the sudden spreading of flame to all remaining combustibles in the space. Personnel still in the compartment when flashover occurs are not likely to survive.

Fully-developed Fire Stage

In the fully-developed fire stage all flammable materials in the compartment have reached their ignition temperature and burning. The rate of combustion will normally be limited by the amount of oxygen available in the air to provide combustion. Flames may emerge from any openings such as hatches, open ventilation ducting, etc. Unburned fuel vapor in the smoke may flash when it meets fresh air in adjacent compartments. A fully developed fire will normally be inaccessible by firefighting teams and require extinguishment by indirect attack. Upon entering into the affected compartment, the firefighting team members must crouch down low because newly fresh air will cause flames once the door to the compartment is open. There may be structural damage to bulkheads or decks when exposed to these extreme temperatures. A compartment may reach the fully-developed fire stage very quickly during machinery space flammable-liquid fires or during enemy weapon-induced fires.

Decay Stage

A fire begins to decay whenever all the available fuels and combustibles in the compartment are consumed. Once all the materials are consumed, the combustion rate slows down until the fire goes out. When the affected compartment does not have sufficient oxygen, the fire may move into the decay stage. The heat release rate decreases as oxygen concentration drops, but the temperature may still continue to rise. The affected compartment may pose a significant because of high concentrations of pyrolyzed fuel, toxic and flammable gases produced by all the combustible materials in the compartment.

Back Draft

If a fire goes out quickly due to a lack of oxygen in a tightly sealed compartment, fuel vapors may still be formed from any flammable liquid that is heated above its flashpoint. If fresh air is allowed into the compartment before this fuel vapor cools below its flashpoint, this mixture can ignite explosively. This process is known as backdraft and the firefighting team has to be very careful when accessing compartments. Wearing their protective gear, crouching low, and using the compartment's door as a shield are used as protection against the explosive ball of fire that will emit once fresh air is introduced to the compartment.

Fire Spread

If the compartment's occupants or rapid responders attack a fire early and efficiently, it can be confined to the area in which it originated and quickly be extinguished. If the fire is neglected, it can generate great amounts of heat that will travel away from the fire area, starting more fires wherever fuel and oxygen are available. Steel bulkheads (walls), decks, and other fire barriers can delay a fire, but not prevent heat transfer. When a fully-developed fire exists in a compartment, the fire will quickly spread to other compartments through openings such as doorways, vent ducts, and unsealed cableways. It will also spread to adjacent compartments by heat conduction through the steel bulkheads. Because heat rises, fires spread faster to the compartment above than to adjacent horizontal spaces. This is why it is extremely important to place boundarymen in the surrounding boundaries but the boundary takes priority. The boundaryman can cool of the deck by placing a ½-inch of water on the deck or wet the bulkhead whenever the deck or bulkhead starts to get hot.

Tests have been developed to provide typical temperatures, radiant heat flux, and length of time for material is ignited by conduction through steel bulkheads from a fully-developed fire. The compartments tested were 8-foot x 8-foot x 8-foot steel cubes with bare metal surfaces. These typical values shown will differ based on the compartment that the fire is affecting will depend on factors such as bulkhead insulation, compartment dimensions, ventilation, specific material characteristics, and water application and cooling.

Fire may spread through bulkhead penetrations such as electrical cableway openings. Although these openings are sealed, experience has shown that even armored cables will burn from extreme heat. Cableway fires may be hard to extinguish because they are difficult to cool because the grouping of multiple cables traps and contains heat. Also, cableways often penetrate the overhead of compartments and heavy smoke hinders finding the source of the fire. Older-style electrical cables will generate toxic black smoke from their insulation, whereas newer cables are designed to reduce the amount of smoke generated.

Firefighting Team Considerations

When a fire is first reported, the fire marshal and rapid response team proceeds directly to the scene to start immediate firefighting efforts. If a fire is beyond the capabilities of the rapid response team, the fire marshal shall order the rapid response team to back out and isolate the affected compartment by closing all surrounding opening. The fire marshal will turn his duties over to the OSL of the Flying Squad in order to coordinate this larger threat. These duties may include the following:

1 Overall command of the Flying Squad.
2 Supervising the establishment and maintenance of communications.
3 Setting boundaries.

The fire marshal will provide all the necessary support to the Flying Squad. The fire marshal assumes a "big picture" role, paying particular attention to the possibility of the fire spreading. The fire marshal will also make recommendations to the DCA for additional personnel needed for the setting of Condition II DC or GQ as required by the size of the fire. These are the steps of succession whenever the ship is out to sea or inport during normal working hours. The IET will respond to fires after working hours and weekends. The base fire department will also respond to assist the IET.

Many factors go into the decision-making process to size up a fire and information is vital. The location, type, and size of the fire, available resources (including personnel), and fire growth all determine the overall plan of attack. Reports from the scene will include the following:

1 Location of fire.
2 Class of fire.
3 Action taken to isolate (electrical isolation, mechanical isolation) and combat the fire (method of attack and activation of installed firefighting system, if applicable).
4 Fire Engaged.
5 No Fire.
6 Reflash watch set.
7 Fire overhauled.
8 Compartment ventilated (desmoking).
9 Compartment atmosphere tested for oxygen, flammable gases, and toxic gases. When CBRN threats exist, reports should include any CBRN contamination and condition of CBRN boundaries.

Much of the initial information about a fire will come from the rapid response team, evacuating watchstanders or initial personnel who took actions to fight the fire. Compartment personnel will evacuate when they are endangered or the fire gets out of control. Every attempt must be made to account for all compartment occupants, since they may not all evacuate through the same exit. All evacuees will meet at a pre-arranged location outside of designated smoke and fire boundaries. Missing personnel must be reported to DCC. Any installed firefighting systems such as AFFF bilge sprinkling, halon, HFP or water mist activation time is documented and reported to the DCRS leader.

Other information may come from the boundarymen or investigators. A boundaryman is responsible for observing a particular bulkhead or deck for signs of heat such as smoldering, blistering paint, or smoke. The boundaryman must especially observe signs of heat coming through bulkhead or deck penetrations. The boundaryman will attempt to cool the bulkhead or deck to prevent spread of the fire as necessary. The investigators travel prearranged routes ensuring that fire and smoke boundaries are set, checking for halon or HFP effectiveness, checking the status of each boundaryman, and making ongoing reports to their associated DCRS. When smoke is encountered, the investigators will immediately report it, put on their SCBAs and conduct further investigation. Various communications circuits are available for the investigators to use to make reports and many ships have hand-held radios to use for firefighting and emergency response situations.

Initial Reports and Compartment Isolation

Information is greatly needed to effectively fight an enormous fire such as a Class B fire in a machinery spaces. This information is gathered by the initial personnel or watchstanders, and passed on to the DCA via the fire marshal, OSL and DCRS leader. The reports given by evacuating personnel will include whether AFFF bilge sprinkling was used, whether the source of the fire (such as leaking or spraying fuel) was secured, and whether halon, HFP or water mist was activated. Also these reports must include an assessment of any damage, trip hazards and any major systems within the affected compartment that were not mechanically and electrically isolated.

For fires in machinery and auxiliary compartments protected by installed firefighting systems such as halon or HFP. Investigators will attempt to determine whether halon or HFP was effective by observing the color of smoke inside the compartment through the battle ports at the escape trunk doors and once they observe this they will make reports to DCRS. Class B fires will produce heavy black smoke and it will diminish in color depending on the effectiveness of halon or HFP. These reports help make the determination whether to immediately re-enter the compartment to combat the fire or wait to allow the space to cool prior to entry.

It is extremely important to know if mechanical and electrical isolation were completed. Electrical power should be isolated before the firefighting team enters the affected compartment because the risk of electrical shock. The decision to secure compartment lighting rests with the OSL, because lighting fixtures are watertight. If vital equipment is needed to be left on during firefighting efforts, the commanding officer will specify this and the firefighting team must be aware and should take precautions near that particular equipment. The Repair Electrician is used to electrical isolate the affected compartment. Mechanical isolation should be completed but does not necessarily need to be completed before sending the firefighting team into the affected compartment; however, it should be in progress. There are great dangers to send the firefighting team into a compartment that fuel systems were not isolated. Firefighting efforts would be pointless because the fuel will continue to feed the fire and the firefighting lives will be at a greater risk. The engineering watchstanders or personnel are used to isolate fuel and oil systems. Also the investigators may be used to help mechanical isolate a compartment.

Fire and Smoke Boundaries

Fire and smoke boundaries are determined for each shipboard compartment. Engineering spaces aboard the ship has to be treated very cautiously because of the many hazards inside of each engineering spaces. The ships have a specific fire doctrine that lists both primary and secondary boundaries for the engineering spaces. The boundaries are designed to effectively contain a fire to prevent its spread. Primary fire and smoke boundaries are set at all bulkheads immediately adjacent to the fire. Boundarymen will manage these primary boundaries with a fire hose and they may have to cool the bulkheads, decks or overheads to prevent the spread of fire. Fire could spread through any penetration, including ventilation, electrical cableways, piping conduits, or defective welds in cases of extreme heat.

A secondary set of boundaries are set at the next immediate watertight bulkhead from the affected compartment. If a boundary fails and the fire cannot be contained at the first boundary, the boundaryman will attempt to secure the space and evacuate. What were previously secondary boundaries now become the primary boundaries. Boundary information is plotted by the DCA and all DCRS leaders.

Preparing to Enter the Space

The ship's *Repair Party Manual* provides information for firefighting and emergency response teams to aid them in successfully control of a casualty situation. This manual provides information in accordance with each ship's configuration and various casualties that may occur to the ship. In particular, the main space fire doctrine is used to aid the firefighting team during a Class B fire in an engineering compartment. This doctrine is a basic checklist that the DCA, DCRS leaders, OSLs, and team leaders must follow while fighting the Class B fire. This doctrine is specifically tailored to each ship.

Briefing the Firefighting Teams

The OSL primary source of information comes from the DCRS leader. The DCRS leader maintains plots of all information about the casualties throughout the ship and will pass along all pertinent information to the OSL. The OSL are responsible for briefing the firefighting and emergency response teams and giving them the necessary information, so they will be better prepared to deal with conditions inside the compartment. Some of the information that the OSL will brief the firefighting teams includes the following:

1 Status of the fire - location, type of fire (and is it still burning), was halon or HFP effective, if applicable.

2 Status of the compartment - extent of major damage, equipment status, mechanical isolation, electrical isolation, boundaries, and known trip hazards.

3 Missing watchstanders or compartment occupants.

4 Activation of AFFF bilge sprinkling or other sprinkling system, if applicable.
5 Planned method of attack.

Dressing Out

Once a casualty is called away the all firefighting and emergency team members must immediately proceed to their assigned DCRS and start dressing out in there protective clothing. Whosever not assigned to the firefighting team and boundarymen will assist the team as necessary to get dressed out. The team leaders must ensure that the firefighting team members are properly dressed out in their FFE. The FFE clothing is designed to fit slightly loose, especially the gloves. This ensures that the skin has room to slightly move inside this clothing. It also helps to keep hotter areas of the clothing from remaining in constant contact with the skin. This practice also reduces the possibility of heat stress by allowing some air movement within the confines of the FFE.

Checking Equipment

When putting on the SCBA, personnel should examine it and ensure it has not been damaged while in storage. The user shall do a quick visual inspection of the face piece, harness straps and cylinder. The face piece should not be cracked or severely scratch to hinder the wearer's vison and the straps should not be deteriorated, unattached, or twisted. The harness straps should not be deteriorated or twisted. The cylinder should not have any major scratches, cracks, or dents. The SCBA air pressure gage should read 4,000 to 4,500 psig. If the SCBA is damaged or pressure is low, turn it into the DCRS leader and get a SCBA that is not damaged and has the correct air pressure.

The team leader shall immediately check the NFTI's operation and battery level and ensure that backup battery packs are charged. The team leader will also ensure that handheld radios and spare battery packs are charged up. The firefighting team members shall ensure helmet lights and voice amplifiers are operational before leaving the DCRS.

SCBA Cylinder Time Tracking

SCBA cylinder time tracking is necessary to maintain as it provides for an uninterrupted attack on the fire by providing hose team reliefs for the firefighting team. The DCRS leader must be able to estimate when SCBA low air alarms will sound. SCBA wearers will report SCBA cylinder start time (and cylinder size if other than 45 minutes) to the DCRS leader via investigators, SCBA coordinator, or the OSL. Reports of SCBA cylinder start time (and cylinder size if needed) can be on an individual basis or on a team basis as needed to support the ship's practice for relieving personnel on scene (either relieving on an individual basis or relieving an entire team). SCBA cylinder start time need not be reported or monitored for individuals, such as investigators, who do not need on-scene reliefs.

Establishing Breathing Air Management

Each ship shall prepare procedures for breathing air management during firefighting evolutions. The extent to which breathing air management functions are implemented depends on the casualty. Breathing air management typically includes setting up an SCBA Recharging Station and/or a SCBA cylinder change-out station. Personnel will provide spare SCBA air cylinders to the SCBA cylinder change-out station, exchange firefighter's depleted air cylinders for full air cylinders, and refill depleted SCBA air cylinders.

The SCBA recharging station is the location of the SCBA recharging connection to the ship's high pressure compressed air system. Stand-alone SCBA breathing air compressor system (SCBA BACS) is an electric compressor installed on ships without high pressure compressed air systems or the portable SCBA breathing air compressor. The DCA is responsible to determine the overall need for SCBA air recharging and to determine which SCBA recharging station(s) will be used. When directed by the DCA, the DCRS leader will activate the SCBA Recharging Station(s).

The SCBA Cylinder change-out station is a designated location where depleted SCBA cylinders can be replaced with fully-charged replacement cylinders. The DCRS leader is responsible for determining the location of the SCBA Cylinder change-out station(s) and allocating manpower. An SCBA coordinator should be identified for every evolution in which SCBA recharging and/or SCBA cylinder change-out is required. If the evolutions are conducted at different sites, more than one coordinator may be needed.

Sizing up the Fire

When sizing up the fire, the OSL and team leader will determine if the size of the fire warrants portable extinguishers, fire hoses, or installed firefighting systems such as AFFF bilge sprinklers, halon 1301, HFP, water mist, or overhead sprinklers. They will also determine best entrance to use to gain access to the affected compartment. Also, they will need to determine what will be the best method to attack the fire. The classification of a fire as well as the location is key information needed when sizing up the fire.

Proper fire boundaries must be set prior to accessing the affected compartment. This provides a safe area from which firefighters can attack the fire. Electrical isolation must be complete prior to re-entry with the only exception is lighting or vital equipment determined by the CO. The OSL will decide whether to secure compartment lighting. Complete electrical isolation helps to decrease the number of ignition sources inside the compartment. Mechanical isolation does not have to be complete prior to re-entry; however, it does provide greater safety for firefighters and isolation should be in progress.

Accessing the Space

If fighting a Main Space Fire (Class B fire) in an engineering compartment, there may be evidence that halon or HFP and bilge sprinkling was not effective prior to compartment re-entry. If secondary halon or HFP is available, it should be used and observed for effectiveness. AFFF bilge sprinkling shall be activated for 2 minutes prior to re-entry. If halon or HFP was effective, allow at least 30 minutes prior to compartment entry. The 30 minutes are broken down into 15 minutes to all halon and HFP to decompose and the last 15 minutes shall be used to remove toxic byproducts of halon or HFP by using installed exhaust ventilation. If halon or HFP was not effective, re-entry should be attempted immediately once mechanical isolation is completed.

A designated hoseman may act as the access person to open the door or hatch so that the firefighting team can enter the compartment. Depending on the size of the hatch or door, two personnel may be designated. If a fire has burned for an extensive time, the access door or hatch to the compartment may be hot and jammed. It may be necessary to use the nozzle to cool off the door or hatch. If the access is still jammed then forcible entry equipment such as bolt cutters, sledge hammers, pry bars, portable exothermic cutting unit (PECU), or portable electrical access and rescue system (PEARS) may be used to gain access.

Hose Team Movements and Hose Handling

When inside the affected compartment, the team leader coordinates the movements of the firefighting team. The coordination of movement is needed in order to move around the affected compartment to extinguish the fire. Sometimes the firefighting team has to move around corners and obstacles. This is when hose handling becomes important in ensuring the hose does not get obstructed which could hinder the firefighting team's progress. Also, positioning on the hose is vital for the firefighting team whenever radio communications fail between the OSL and team leader. When radio communications fail, information can be passed and received by the OSL and team leader utilizing each firefighting team member sending the message back-and-forth to amongst the hose team.

The first obstacle for firefighting team members is a ladder leading downward. Because heat can weaken ladder rungs, safety must be observed by ensuring that the first hose team member checks the strength of each ladder rung as they ascend and only one person should be on the ladder at a time. When bringing a charged hose down an inclined ladder, one method is to carry the hose over one's shoulder with the nozzle in front in one hand and the other hand on the rail. When bringing a charged hose down a vertical ladder, the nozzleman will climb down the ladder while the nozzle and hose is position in between the nozzleman and ladder. The nozzle and hose is lowered as the nozzleman descends down the ladder. Proper hose tending by the hosemen will ensure that the nozzleman on the ladder maintains his balance. As the nozzleman advances, the hose team members pass the hose down to him while he descends the ladder. After he reaches the deck, the first hoseman will descend the ladder, followed by another hoseman, as needed to handle the hose. As the hose progresses further into the compartment, more hose is needed, as well as hosemen (Figure 6-1).

The team leader usually operates the NFTI, looking for hotspots and hidden fires. Although the team leader already knows the location of the seat of the fire, they must be alert to the likelihood that other parts of the compartment are on fire. The team leader must also look for obstructions that prevent advancing to the seat of the fire. The team leader will also issue orders for hose advancement and instructing the nozzleman to attack the fire with the necessary spray pattern.

The nozzleman should devote one hand to holding the nozzle and directing the stream. The other hand should be available to alternate between the pattern shroud and the bail shut-off handle to control pattern or flow variations as appropriate. This hand, knuckles up, on top of the nozzle is typically the most comfortable and provides for the least movement between pattern shroud and bail handle. When using a tactic of short bursts of agent, this hand would typically be kept on the bail. For continuous flow tactics, keep the hand on the shroud for instant pattern variations and to assist in aiming the nozzle. On the 1-1/2 inch vari-nozzle, the pistol grip handle may or may not be used, whichever methods is most comfortable and works best for the nozzleman. The nozzle may be held with the hose over the shoulder, under the arm, across the knees, or other method which works best to provide the necessary control.

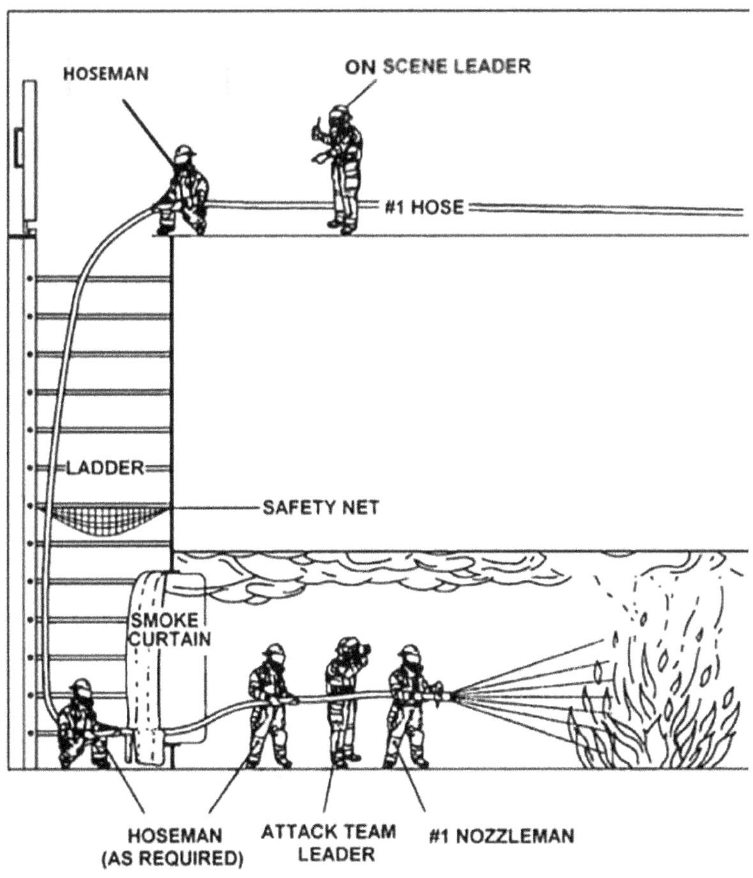

Figure 6-1. Accessing a Fire down a Ladder

Hosemen follow the direction of the team leader, moving forward on the hose, advancing or backing up with the hose, and handling the weight of the hose. Whenever the nozzle is opened, a recoil effect pushes the hose backwards, and hosemen must push forward to compensate for this. The fire hose is easiest to advance when the nozzlemen carries the nozzle, assisted by their backup hoseman, and additional hosemen carrying each coupling. Additional personnel, if available, should be located at the mid-length of each 50-foot section of hose and at turns which the hose makes. This procedure reduces friction due to dragging on the deck and prevents couplings from being snagged on door openings or other obstructions.

Attacking the Fire

The attack should begin as soon as possible to gain immediate control and to prevent or minimize the spread of fire. During fire size up, the OSL and team leader must evaluate the fire conditions of the affected compartment and select the appropriate tactic to control and extinguish the fire. If access to the affected compartment is available, a direct attack should be conducted. Direct attack options include direct attack on the seat of the fire, fog attack, or direct attack from the access. If conditions in the affected compartment prevent access, actions should be taken to improve conditions to permit a direct attack by conducting an indirect attack and venting the fire in the affected compartment.

Direct Attack

The most common method to fight a fire is a direct attack. During a direct attack, the firefighting team proceeds to the immediate fire location and apply water directly onto the fire. The team leader will determine which technique to use. There are three distinct techniques that can be used during a direct fire attack as follows:

1. Direct attack at the seat of the fire.
2. Fog attack (to gain control of fire).
3. Direct attack from the access.

If conditions become too severe for these direct attack techniques, the firefighting team must withdraw and use other techniques such as indirect attack or venting to improve conditions to allow a direct attack.

Direct Attack at the Seat of the Fire

When direct access to the seat of the fire is available, the preferred method is to advance towards the fire and apply water directly onto the seat of the fire for extinguishment (Figure 6-2). Access to the fire may be straight forward in the early stage of the fire, but heat, gases, and smoke from an advanced fire make access increasingly difficult. The direct attack technique involves application of short bursts (several seconds) of water with a narrow angle fog or straight stream nozzle pattern onto the seat of the fire. The nozzleman pauses and lets the resultant steam pass over and subside. During this pause, the firefighter listens for noise to help locate the fire. After the steam has subsided, water is again applied in a short burst. The duration of the short burst is a function of the amount of steam production. As steam production lessens, a greater duration of a water burst may be used as the firefighting team moves forward and attempts to directly attack the fire.

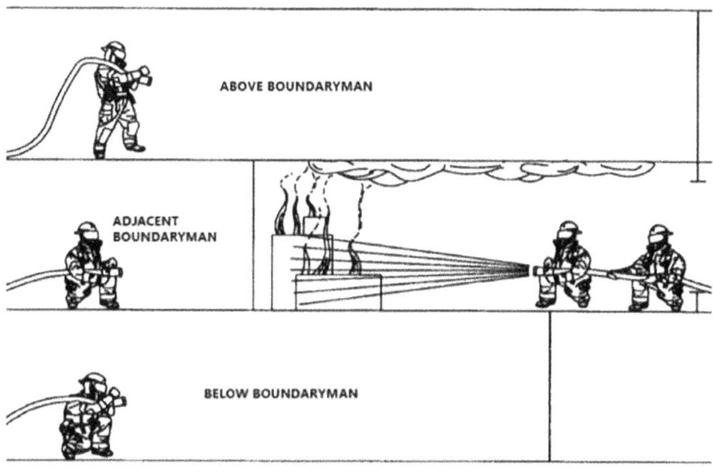

Figure 6-2. Direct Attack

The practice of using a continuous water flow during fire when making an interior entry should be discouraged. A continuous flow of water should only be used when the firefighting team is faced with the fire or when the team cannot approach closer due to radiant heat. This does nothing more than disturb the thermal balance in the compartment which will produce large amounts of steam before the firefighting team has a chance to advance on the fire. Also the steam will reduce visibility to the point where the firefighting team cannot see where they are going or located the source of the fire. An issue that may be created when using a continuous flow of water is potential flooding. Seawater adds weight to the ship and if this potential flooding occurs on the top levels of the ship then there may be stability problems that potential put the ship at risk of capsizing. The technique of using short burst should be used by the firefighting team until they are directly at the seat of the fire. Short burst should also be used when cooling a fire boundary as well.

Fog Attack

When entry can be made into the affected compartment, but direct access to the seat of the fire is not possible, the firefighting team may use a fog attack to gain control of the fire and delay or prevent flashover (Figure 6-3). The following conditions indicate when to use a fog attack to gain control:

1. Where overhead gases are burning (known as rollover).
2. Where the seat of the fire is obstructed and water streams cannot be applied directly to the seat.
3. Where multiple seats of the fire are growing within a compartment such that one seat of the fire would grow out of control while water is being applied to another seat of the fire.

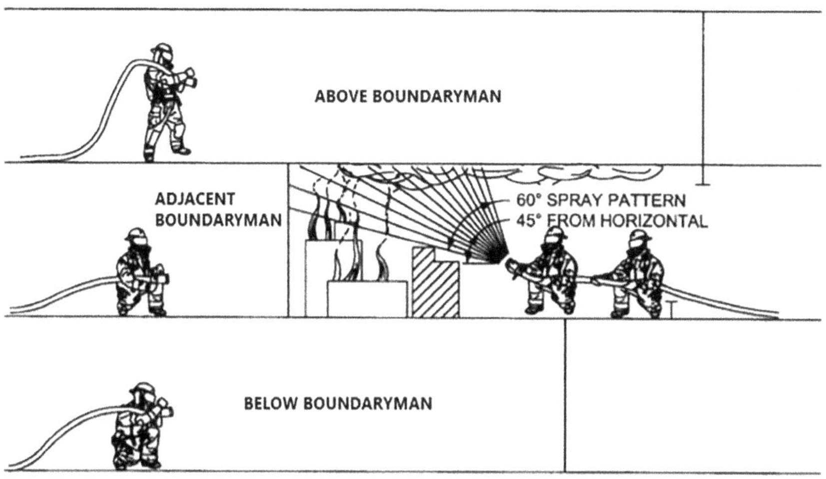

Figure 6-3. Fog Attack

Fog bursts applied directly over the seat(s) of the fire will reduce the affected compartment temperatures, radiant heat, and flaming combustion. When conducting a fog attack, the nozzleman will position the vari-nozzle to the medium angle (60 degree) pattern and apply water in short bursts aiming the nozzle approximately 45 degrees towards the overhead (ceiling) gas layer to control the fire. The nozzleman will fully open the nozzle to apply a short fog burst (2 to 3 seconds) into the upper gas layer followed by a 2 to 3 second pause. During the pause, the team leader checks for flames and directs additional fog bursts until the flames are knocked down. Use of short fog bursts followed by short pauses will maintain acceptable conditions in the compartment for the firefighting team. If the fog bursts are too long, the wrong nozzle pattern is selected, or the nozzle is not directed properly, then too much water may be applied onto the hot steel surface of the overhead or bulkhead creating excessive steam. Excessive steam could force the firefighting team to withdraw. During a fog attack, the firefighting team must use the Navy vari-nozzle and never use the Navy all-purpose nozzle. Once flames are knocked down, the firefighting team must immediately advance towards the seat of the fire to conduct a direct attack.

Direct Attack from Access

If high temperatures deny a direct access into the affected compartment but the burning material can be reached by using a hose stream from the compartment's access, then the nozzleman can apply water to the seat of the fire directly from the entrance (Figure 6-4). The nozzleman may use the bulkhead at the access as a shield. If the seat of the fire can be reached by a hose stream, then the nozzle should be set to straight stream or narrow angle fog directed at the seat of the fire. A smoke curtain can also be hung at the door to shield the firefighting team. If the seat of the fire cannot be extinguished by direct attack, a fog attack may be used to reduce the compartment's temperatures and knock down flaming combustion. When conditions permit, the firefighting team must advance into the immediate fire area and apply water directly onto the seat of the fire.

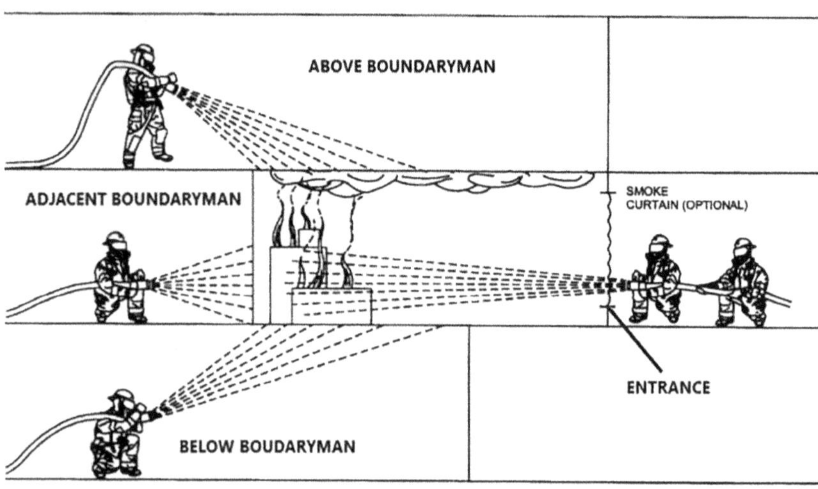

Figure 6-4. Direct Attack from Access

Indirect Attack

An indirect attack is the application of water fog into the fire affected compartment through an existing access or through a hole cut in a bulkhead or overhead. The goal is to distribute water throughout the hot upper layer in the affected compartment and generate enough steam to effect extinguishment, to gain control of the fire, or to reduce compartment temperatures. Water is preferred over AFFF because it will create steam and absorb heat more effectively than AFFF. When heat or other conditions deny access to the affected compartment, an indirect attack may improve conditions to enable the firefighting team to enter the compartment for a direct attack on the fire. An indirect attack is mostly effective in extinguishing the open flames of a Class A and Class B fire in an enclosed compartment when the physical boundaries are not breached by major openings. Major openings in the ship's hull can occur from weapons or collisions which may have caused the fire.

The basic technique for an indirect attack is to apply water fog to the affected compartment for approximately five to ten minutes. Once the prescribed time has elapsed, the water flow is stopped while the conditions of the fire are assessed through a cracked access. The fire conditions can be assessed by using a NFTI to monitor temperatures or by the severity of heat and steam that blows out of the access from the affected compartment. If it appears that the compartment can be entered, the firefighting team must immediately enter the compartment to conduct a direct attack on the fire. If the compartment cannot be entered, then indirect attack and assessment of the fire conditions will continue until the affected compartment can be entered. The team leader should adjust the application times and indirect attack points based on their assessment of the fire conditions. An indirect attack shall never be conducted whenever the firefighting team occupies the affected compartment or whenever a direct attack is in progress. An indirect attack will generate steam and hot water droplets will create a moderate hazard to the firefighting team if they are in the affected compartment. The affected compartment must be isolated during the indirect attack. It is extremely important to keep the affected compartment isolated between indirect attacks to prevent injury to personnel. Also ventilation within the affected compartment shall be secured during indirect attack to assist in retaining steam formed in the compartment and reducing oxygen concentration for fire extinguishment. Normal ventilation should be secured in the compartment which the indirect attack is staged unless active desmoking or positive pressure has been established.

When conducting an indirect attack, a Navy all-purpose nozzle with 4-foot applicator with the large orifice nozzle is effective during the process, but a vari-nozzle with wide angle fog pattern may be preferred for indirect attack. The Navy vari-nozzle is preferred because it will allow the firefighting team to quickly enter the affected compartment to conduct a direct attack on the fire. If the location of the fire is known, the firefighting team can conduct an indirect attack from a position that allows application of water to the fire. If there are obstructions around the fire and the affected compartment is large, it may be necessary to conduct simultaneous indirect attacks from multiple points utilizing two firefighting teams. If the fire is spread out or its location is uncertain, separate the attack points to maximize coverage of the affected compartment by water spray from multiple firefighting teams. If multiple teams are utilized communication and coordination are extremely important between teams.

Indirect Attack through an Access

Indirect attack can be conducted through several accesses to the affected compartment by a door, hatch, or scuttle. Indirect attack is conducted through a door, scuttle or hatch by cracking the closure enough to place the nozzle through it. The steps through a hatch or scuttle are as follows:

1. Crack open the hatch or scuttle just enough to admit the nozzle tip or extend the applicator such that the applicator head is clear of obstructions.
2. Once the nozzle or applicator is in place, lower the scuttle or hatch down onto the nozzle or applicator. The nozzleman can slide their hands up the hose away from the nozzle to reduce exposure to steam while still maintaining control of the nozzle.

When conducting and indirect attack through a door utilizing a vari-nozzle, the top of the door sill is normally located several feet below the compartment overhead and will usually have better heat and steam conditions than a hatch or scuttle. The steps through a door using a vari-nozzle are as follows:

1 Set a vari-nozzle on medium-angle spray.
2 Open the door with the vari-nozzle located at an appropriate distance (roughly two feet) such that the initial spray fills most of the door width.
3 Initiate an up and down nozzle motion to establish a venturi effect in the door. As conditions permit, move forward to the door and rotate the nozzle in a circular motion to distribute the water spray through the compartment (Figure 6-5). The nozzle can be extended up to an arm's length through the door. An applicator is expected to provide less protection for the team from steam and heat kick-back from the compartment.

Figure 6-5. Indirect Attack

Indirect through a Hole in Bulkhead or Overhead

The Navy all-purpose nozzle with 4-foot applicator is effective when indirect attack is utilized through a hole cut through the bulkhead or deck using a PECU. The size of the hole should only be cut to enough clearance to allow the 4-foot applicator to be placed into the hole. The steps during this process as follows:

1. Cut a hole in a bulkhead or in the overhead of the affected compartment.
2. Insert the applicator into the hole and apply water fog into the affected compartment continuously for approximately five to ten minutes.
3. Secure the water flow and conduct an assessment of the fire conditions through a cracked open access. Based on the assessment, the firefighting team shall enter the affected compartment and conduct a direct attack or conduct another indirect attack until favorable conditions are met.

It may be necessary to leave the applicator in place in case indirect attack is needed again. The applicator can be left in placed unmanned but if this occurs it shall be properly mounted and constantly monitored. This process will only be recommended if steam and heat conditions present a hazard to personnel. Compared to using a cracked open access for the indirect attack, attacking through a cut hole in a bulkhead or overhead will reduce the amount of heat and steam blowing out of the affected compartment. The more the steam is kept in the affected compartment the more effective the indirect attack.

Completion of Indirect Attack

Maintain the water spray until steam no longer blows from the access and until water has been applied for approximately five to ten minutes. Since the principal function of indirect attack is to reduce the temperature and improve conditions of the affected compartment, it is important that the water application be continued for a reasonable period of time to remove the energy from the affected compartment. If steam forces the firefighting team back during the indirect attack or after it is completed, the firefighting team will close the access and wait for one or two minutes. After the one or two-minute period, the firefighting team will crack open the access to assess the conditions. After the quick assessment, the OSL will decide to enter the affected compartment to conduct a direct attack or conduct another indirect attack until favorable conditions are presented.

Compartment Venting

Compartment venting is another means of cooling the space so firefighters may safely enter the affected compartment. When deciding to cut holes above the fire to vent heat to the exterior of the ship, large areas open to the weather decks such as a hangar bays, flight decks, or well decks should be considered and an indirect application of water can be applied to the affected compartment. If heat can be vented from the overhead of the fire compartment, the resultant temperature reduction and smoke removal will permit firefighting teams to advance directly toward the fire. If there is a weather deck directly above any portion of the affected compartment, heat venting should be attempted by cutting holes at least a one-foot square in the deck. The larger the hole the quicker the heat and smoke will be vented from below. Similarly, vari-nozzles or applicators can be inserted in holes cut into the affected compartment. For safety, the personnel who are cutting holes must be in full protective clothing such as the firefighter's ensemble, SCBA, and SCBA face piece equipped with a welding lens. Holes should only be cut from weather decks or large areas open to the weather decks such as hangar bays, flight decks, and well decks to prevent personnel getting engulfed by heat, smoke, and steam that rapidly escapes from the vent hole. Wind direction across the deck must be considered to prevent heat and smoke from affecting other vital activities and vital control areas such as the bridge. The DCA may request the OOD to position the ship for favorable wind conditions.

Inside the Affected Compartment

All the different fire attacks have been covered and no single tactic or strategy is applicable to every situation. The indirect attacks and fog attacks are used to enable the firefighting team to get to the seat of the fire. Once the firefighting team gets to the seat of the fire the ideal method of attacking the fire is a direct attack. This technique involves short bursts with a narrow fog or direct stream, as directed by the team leader. In extreme high temperature conditions, deck plates or ladders steps may have become weaken so the firefighting team must move throughout the compartment or down the ladder with extreme caution.

Extinguishment

Once firefighting team has successfully reached the seat of the fire, the team leader makes a report to the OSL that the fire is engaged. The team leader will continue to direct the nozzleman to extinguish any remaining fire. Different spray patterns from the hose nozzle may be used as needed, either to break up any combustible material, or to cover a certain area with AFFF if the fire is a Class B. To reduce the potential of heat stress and fatigued the team leader will order the hosemen to periodically relieve the nozzleman. The firefighting team will continue to fight the fire until it is extinguished. The firefighting team will get relieved once the team's SCBA breathing time is approximately 15 minutes remaining, but may be early as 20 minutes.

Prevention of Reflash

AFFF is highly effective against Class B fires, because it serves three distinct functions. As foam it floats on top of flammable liquids, preventing vapors from being released to the atmosphere. This foam also prevents oxygen from reaching the flammable liquid. The AFFF foam, being a mixture of concentrate and water, also provides a cooling effect. Therefore, covering hot spots with AFFF is highly effective in preventing reflash. Allowing the compartment to cool down after halon or HFP has been effectively used will help prevent reflash once the firefighting team enters the compartment.

Reflash Watch

Once satisfied that the fire is extinguished, the team leader directs one of the firefighting team to set the reflash watch. The person assigned as reflash watch remains near the seat of the fire with a charged hose and observes the area to ensure that no new fire breaks out. At least one hoseman remains with the reflash watch to tend the hose in case the fire reflashes.

Overhaul

Once the reflash watch is set, the team leader and other members of the firefighting begin the overhaul process. Overhaul of a fire is an examination and cleanup operation which includes finding and extinguishing hidden fires and smoldering materials. During overhaul, if any hidden fires are found they are extinguished and if any smoldering materials are found they are cooled. It is best to start the overhaul process in the affected compartment at the perimeter of the fire and work towards the origin of fire. All areas of the compartment are examined with the NFTI ensuring that no areas are missed. All cableways, areas beneath deck plates, and overheads are examined to ensure no hidden fires are missed. Check for all possible areas of fire spread, including behind electrical outlets or cables, inside vent ducts, in thermal or acoustic insulation, and in concealed spaces. Clues to concealed fires are smoke creeping out of openings, surfaces hot to the touch, and results from the NFTI. These clues can be used to trace hidden fire. It is sometimes necessary to use overhaul equipment such as a rake to pull down smoldering or burning material, bulkhead insulation, false overhead, and all remaining fire material (such as lagging) in order to extinguish it. At various times, the team leader will make reports detailing percentage of overhaul.

Desmoking

After the fire is extinguished, smoke and combustible gases may be present. This is the reason for post-fire desmoking. For all classes of fire, carbon monoxide will be the predominant gas. Although combustible, substantial quantities of carbon monoxide must be generated to reach the flammable range (12.5 percent is the lower flammable limit). Heavy carbon monoxide production which does not ignite is typically associated with a confined fire that has smoldered for several hours. Smoke and combustible gases can be removed using potable blowers such as the Ramfan, box fan, or installed ventilation fans.

For a compartment which has experienced a Class A or Class C fire, after the fire is out (visible open flaming has been extinguished), desmoking to support fire overhaul can proceed with minimal risk. There are no recorded incidents where a post-fire atmosphere involving Class A or Class C compartment fires has ignited as a result of operating installed ship's ventilation systems. When a Class B fire has been extinguished, combustible gases may be present. Operating electric controllers to start ventilation fans may ignite these gases. Desmoking with installed ventilation can proceed with minimal risk once specific conditions are met. These conditions include the following:

1. The fire is extinguished and overhauled.
2. The AFFF bilge sprinkling has been operated.
3. The source of the fuel for the fire is secured.
4. The compartment has been allowed to cool.
5. All fuel has been washed to the bilges.
6. No damage has been sustained to the electrical distribution system.

Desmoking should begin once the compartment has cooled sufficiently so there is no danger from re-ignition. Circuit breakers that have tripped should not be reset until qualified personnel can make a damage assessment. Examine the electrical distribution system, and if possible, reestablish power to the installed ventilation fans. If the fans are fully operational, run them on high speed for a minimum of

15 minutes to remove smoke and toxic gases. If the installed system is partially operational or inoperative, desmoking will take longer, but can be accomplished by using portable blowers or by providing a positive ventilation from adjacent compartments. The goal in desmoking is to replace 95 percent of the smoke-laden air with fresh air. This will require approximately four complete space volume changes in a compartment. This can be achieved with a fully operable ventilation system in 15 minutes for flammable liquid storerooms and issue rooms, machinery rooms, auxiliary machinery rooms, fuel pump rooms, and diesel generator rooms. Ship's ventilation drawings must be consulted in order to calculate minimum ventilation times for other compartments. On ships without halon or AFFF bilge sprinkling, the safest method of desmoking is to exhaust the compartment with portable fans, or to provide a positive ventilation pressure from adjacent compartments.

Atmospheric Testing

Atmospheric tests are always conducted after desmoking is complete because combustible gas indicators will not operate reliably in a halon or HFP atmosphere and an oxygen analyzer is unreliable when its sensor is exposed to excess moisture or comes in contact with particulates found in a post-fire atmosphere. The testing is conducted by the GFE, GFEA, or GFEPO. When the compartment is clear of smoke, the atmosphere is tested for oxygen, combustible gases, and toxic gases. The level of oxygen must be between 19.5 and 22 percent. Combustible gases must be less than 10 percent of the lower explosive limit and all toxic gases must be below their threshold limits before the compartment is certified safe for personnel without breathing apparatuses.

During a Class A or C fire the toxic gas that is tested will be carbon monoxide (CO) and carbon dioxide (CO_2). When polyvinyl chloride electric cable jacketing is burned, the toxic gas that shall be tested is hydrogen chloride (HCl). When vinyl nitrile rubber chilled water piping insulation is burned, hydrogen cyanide (HCN) is produced. An additional toxic gas test for hydrocarbons is required for Class B fires. Hydrogen fluoride

(HF) is produce from decomposition of halon 1301 or HFP and these gases shall be tested also. The required tests shall be conducted near the center and at all four corners of the compartment. If the compartment has multiple levels the required test shall be conducted on each level of the compartment. At least one satisfactory reading at each location must be obtained. These readings will be given to the GFE to help determine when breathing apparatus can be removed.

Dewatering

When firefighting is being conducted is a great chance that water is left in the affected compartment. Firefighting water can add weight to the ship which can decrease the ship's stability and buoyancy. Dewatering is removing flooding water or hazardous liquid from the ship by using dewatering equipment such as portable eductors, P-100 pump, electrically submersible pumps and installed eductors. Before dewatering is conducted the commanding officer must give the DCA permission to dewater in accordance with operating procedures. Dewatering a Class B pool fire will not commence until the affected compartment has been completely overhauled, except in extreme conditions where ship's stability is threatened. Dewatering will affect the vapor barrier on top of pooled flammable liquid, so an extreme caution must be exercised to ensure the AFFF blanket is maintained until completion of overhaul. Following overhaul, normal dewatering may be conducted or completed at the same time as desmoking or post-fire gas free testing.

Post-Fire Investigation

After overhaul, the fire should be investigated to determine the point of origin, types of combustibles involved, path of fire spread, ignition source, and significant events in the growth and eventual extinguishment of the fire. Starting from the point of farthest fire spread, burn patterns will usually extend back to the area of origin. Efforts should be directed toward recreating the conditions that caused the fire and identifying any changes in design or procedures that could have prevented the fire or lessened its spread and intensity. These changes are very helpful to ship designers and operators. Also information from the investigation is helpful to improve firefighting procedures, systems and equipment. Photographs, material samples, metallurgical samples, and failed equipment assist in reconstructing a fire history. If there is a major fire which involves significant damage or loss of life, a Naval Sea System Command (NAVSEA) technical expertise team is available to investigate such fires and to develop lessons learned from a ship design and a material standpoint.

This chapter provides information pertaining to the tactics and strategies involved in fire- fighting. While every fire is different, certain practices will apply to all fires. This chapter was a basic snapshot of firefighting tactics more detailed information on extinguishing different types of fires found in Naval Ships Technical Manuals. While the information is located in these manuals, there is no substitute for actual hands-on training. The fire protection engineer will train shipboard personnel in firefighting, as well as other aspects of damage control. Shipboard personnel can become proficient as a firefighter by engaging in drill scenarios onboard ships and attending both the General Shipboard Firefighting and Advanced Shipboard Firefighting courses held at various locations in the U.S. States and overseas. A properly trained firefighting team may make the difference between dealing with a small easily controlled fire and an enormous one that threatens the entire ship. The next chapter shall cover the major players in the aviation firefighting organization.

CHAPTER 7

Aviation Fire Protection Engineering

Fire protection is critical onboard air-capable ships during flight and maintenance operations. The degree of fire protection will be based on potential hazards involved during flight evolutions. As mention in the first chapter, the fire protection engineer in the aviation domain is known as the aviation boatswain's mates. The aviation boatswain mate is a military occupation specialty (MOS) that consists of three specialized occupations: equipment, fuels and handling. They play a major part in launching and recovering naval aircrafts quickly and safely from land or ships. This includes aircraft handling, recovering, refueling, firefighting, rescuing, and crash salvaging. As an aviation fire protection engineer, they will perform preventive and corrective maintenance to all fire protection system, fire protection equipment, fire protection vehicles, and rescue equipment related to aircrafts, flight decks, and hangar bays onboard air-capable ships. They hold key positions on the firefighting, crash, rescue and salvage teams and they train other aviation occupations in specialized techniques used for aircraft firefighting.

Key Roles and Responsibilities

An effective fire protection program in the aviation community is very vital to saving lives, protecting flight decks, hangar bays and its aircrafts. The success of the firefighting organization begins with proficient and dynamic leadership. The key personnel and positions that play major roles on air-capable ships are as follows:

1. Air officer (Air Boss).
2. Helicopter control officer (HCO).
3. Aircraft handling officer (ACHO).
4. Aircraft crash, salvage, and rescue officer (Air Boatswain).
5. Aircraft crash, salvage, and rescue supervisor.
6. Hangar deck officer (HDO).
7. Aviation fuels officer.
8. Integrity watch officer (IWO).
9. Ordnance handling officer (OHO).
10. Air gunner.
11. Air department training team.

Air Officer (Air Boss)

The air officer, also known as the air boss, is responsible to the ship's commanding officer for the supervision and direction for aircraft launching, landing, and handling operations. The air boss is also responsible for following:

1. Operation, inspection, and maintenance of aircraft handling equipment which includes elevators, tractors, tow bars, firefighting vehicles.
2. Operation, inspection, and maintenance of catapults, arresting gears, and visual aids.
3. The care, stowage and issue of aviation fuels and lubricants.

4 Maintenance and inspection of all firefighting systems and equipment related to the flight deck and hangar bays.
5 Aircraft firefighting, salvage, jettison, and personnel rescue.
6 Coordination with the DCA to combat fires in hangar bays.
7 Order the cessation of all fueling/pumping evolutions.
8 Order the activation of applicable conflagration stations for the hangar bay or AFFF zones for the flight deck as necessary.
9 Plan to move aircraft away from the fire, including provisions for wetting down (cooling) aircraft that have been subjected to high temperatures.
10 Develop contingency plans for fighting fires involving ordnance.
11 Prepare other contingency plans for maintaining a ready deck area clear of hazards for operation of aircrafts assisting in emergency operations.

Helicopter Control Officer (HCO)

The helicopter control officer is an assistant to the air boss on LPD class ships. On other air-capable ships other than the LPD, the HCO is normally held by a junior officer. The responsibilities of the HCO shall include the following:

1 Aircraft firefighting, salvage, jettison, and personnel rescue operations.
2 Assesses the fire situation, advises the CO, and requests assistance commensurate with the gravity of the incident.
3 On other air-capable ships, directs all aircraft firefighting and rescue operations on the flight deck and vicinity under the supervision of the DCA.

Aircraft Handling Officer (ACHO)

The aircraft handling officer (ACHO) is one of assistants to the air boss. The ACHO is responsible for the movement of the aircrafts on the flight deck and hangar bays. During fires and crashes the ACHO shall be responsible for coordinating the movement of all aircrafts and ensuring communications are established between the OSL and primary flight control.

Aircraft Crash, Salvage, and Rescue Officer (Air Boatswain)

The aircraft crash, salvage, and rescue officer is known as the air boatswain. The air boatswain is an assistant to the air boss and this position is usually held by a chief warrant officer, a subject matter expert of the aviation boatswain's mate MOS. The air boatswain is responsible for organizing, supervising, and training the crash, salvage, and rescue teams for both flight and hangar bays. Also the air boatswain shall be responsible for the operation and maintenance of assigned equipment on the flight deck.

Aircraft Crash, Rescue, and Salvage Supervisor

The aircraft crash, rescue, and salvage supervisor is a position that is used only on LPD class ships. They perform the same role as the air boatswain. This position is usually held by a chief petty officer or a subject matter expert of the aviation boatswain's mates MOS. The aircraft crash, rescue, and salvage supervisor is responsible for organizing, supervising, and training the crash, rescue, and salvage team as well as the operation and maintenance of assigned equipment on the flight deck.

Hangar Deck Officer (HDO)

The hanger deck officer is an assistant to the air boss. The HDO is responsible for all operations and maintenance of firefighting systems and equipment located in hangar bays. During firefighting evolutions, the HDO shall supervise the applicable conflagration station and order the activation of the appropriate AFFF overhead sprinkling zone.

Aviation Fuels Officer

The aviation fuels officer is an assistant to the air boss. They are responsible for the organization, training, and operation of the aviation fuel repair team. Their specific duties shall include:

1 Maintenance and operation of the aviation fuel system and equipment.
2 Ensure safe refueling and defueling evolutions.
3 Provide personnel for the background assistance team during crashes and fires.
4 Ensure personnel are qualified in firefighting and rescue positions.

Integrity Watch Officer (IWO)

The integrity watch officer is a vital watch position that is set after flight operations. This watch is stood by various aviation personnel. The IWO roves around the flight deck and hanger bay ensuring that there is no presence of hazards to the aircrafts and ship such as fuel leaks or fires. Their specific duties during emergency conditions shall include:

1 Immediately report hazard to flight deck control.
2 Apply initial emergency response actions.
3 Move aircrafts to safety.

Ordnance Handling Officer (OHO)

The ordnance handling officer is attached to the weapons department. The OHO is used on CVN class ships. They shall inform the ACHO and air boatswain whenever dealing with ordnance on the flight deck and hangar bay. The will specifically report the following:

1 The type of ordnance.
2 The location of the ordnance.
3 Keeps the ACHO and air boatswain updated on the status of the ordnance.

Air Gunner

The air gunner sometimes referred as the gun boss is attached to the weapons department. They perform the same duties as the CVN class ships' OHO. They shall inform the ACHO and air boatswain whenever dealing with ordnance on the flight deck and hangar bay. The will specifically report the following:

1 The type of ordnance.
2 The location of the ordnance.
3 Keeps the ACHO and air boatswain updated on the status of the ordnance.

Air Department Training Team (ADTT)

The air department training team (ADTT) consists of subject matter experts with strong backgrounds in crash, salvage, and rescue evolutions. This team will be responsible for training the crash and salvage team through on-the-job training and drills. Air-capable ships that do not have an air department will have a designated aviation training team (ATT). ADTT may integrate with other shipboard training teams such as DCTT to implement drill scenarios that are challenging to the crash and salvage team. Challenging scenarios will help develop an effective team to be able to respond to real life events.

Aviation Firefighting and Emergency Response Teams

During flight operations there will always be risk of crashes and fires occurring on flight deck and hangar deck. The degree of fire protection readiness must be at the pinnacle level. Firefighting teams must be able to commence firefighting and rescue evolutions at instantaneously. The firefighting and emergency response organizations onboard the CVN, LHD, LHA, and LPD class ships are known as the following:

1 Crash, salvage, and rescue team.
2 Hangar deck firefighting and rescue team.
3 Aviation fuel repair team.

Crash, Salvage, and Rescue Team

The crash, salvage, and rescue team also called the crash and salvage team responds to all crashes and fires on the flight deck. This team is made up of various personnel that have specific roles during fires and crashes. Their specific duties are as follows:

1 Rescue of personnel from damaged aircrafts.
2 Rescue injured personnel on the flight deck.
3 Clear away wreckage on the flight deck.
4 Make minor emergency repairs to the flight deck and its associated equipment.

Hangar Deck Firefighting and Rescue Team

The hangar deck firefighting and rescue team functions same as the crash, salvage, and rescue team. This team responds to all fires in the hanger deck. This team is made up of various personnel that have specific roles during fires.

Aviation Fuel Repair Team

The aviation fuel repair team conducts fueling operations. This team consists of personnel in the aviation fuel specialty. This team specifically functions in the following manner:

1 Respond as the background assistance detail during crashes and fires.
2 Conducts emergency repairs to the fuel systems and equipment.
3 Get dispatched to isolate affected fueling stations and quadrants of the JP-5 system and make reports to flight deck control once systems are isolated.
4 Performs firefighting evolutions on fires affecting the fueling stations.

Aircraft Firefighting Team Positions

The aircraft firefighting team organization and duties are essential to support all air-capable ships' mission. The training and qualification process is the key link to fostering an effective firefighting and emergency response team. All the members of a well-organized crash and salvage team should be cross-trained in all aspects of crash and salvage. Each member should be trained and qualified in any assigned positions. The following positions are part of the crash and salvage team:

1 On-ene leader.
2 Hose team leaders.
3 Hose team members.
4 AFFF station operators (plugman).
5 AFFF proportioner station operator.
6 Rescue personnel.
7 Overhaul personnel.
8 Background assistance leader.
9 Background assistance detail.
10 Messenger.

11 Medical personnel.
12 Weapons personnel.
13 Mobile firefighting vehicle (MFFV) driver and operators.
14 Conflagration (CONFLAG) station operator.

Scene Leader

The scene leader has similar responsibilities as the OSL during shipboard firefighting evolutions. The scene leader will assume command at the location of crashes and fires. The scene leader will direct personnel to commence firefighting procedures and tactics, ordnance cooling, and personnel rescue as required. The scene leader shall do the following:

1 Immediately make an appraisal in regard to the presence of hazardous ordnance and request confirmation from flight deck control.
2 Determine if additional hose teams are needed.
3 Direct firefighting teams in weapons cooling.
4 Evaluate the fire and make recommendations to flight deck control for maneuvering the ship to provide favorable wind conditions.
5 Ensure hose teams attack the fire from a 45-degree angle, when able.
6 Maintain visual contact for hand signal communications with hose team leaders, rescue, and overhaul personnel.

Hose Team Leaders

The hose team leaders have similar responsibilities as the hose team leaders during shipboard firefighting evolutions. The team leaders position themselves behind the nozzleman of the hose teams. The team leaders follow direction from the scene leader and direct all actions of the hose teams during firefighting evolutions. The hose team leaders have the following responsibilities:

1 Ensure the hose teams are manned.

2 Ensure the AFFF hose or fire hose is laid out.
3 Directs the hose team to position themselves on the outside hose adjacent to the fire.
4 Directs the AFFF station operator or plugman to charge the AFFF hose or fire hose.
5 Directs the nozzleman to test for firefighting agent and nozzle patterns.
6 Directs the hose team movement and when applicable, maintain a 45-degree angle in relation to the fire.
7 Directs hose team to maintain distance and cool ordnance, when applicable.
8 Maintain visuals with scene leader, use hand signals or verbal to report fire status.

Hose Team Members

Hose teams perform similar functions as the shipboard firefighting hose teams. The hose teams should be positioned on the outside of the hose in relation to the aircraft to aid in mobility, communication, and decrease interference between hose team members when applicable. The hose teams will use a AFFF hose when fighting aircraft and fuel fires and will use a navy standard fire hose when cooling ordnance. The number of hose team members will determine on the size of the fire hose (1 ½-inch or 2 ½ inch). Hose teams shall take the following actions:

1 Maintain a stance in a position that is the safest and most effective to fight the fire.
2 Maintain a low stance to avoid flames, heat, explosions, and smoke.
3 Attach the fire and maintain a 45-degree angle, if applicable.
4 When applicable commence weapons cooling.
5 Follow directions of the team leaders.
6 Maintain verbal communication back-and-forth with the team leader and scene leader.
7 Maintain reflash watch during post fire procedures.

AFFF Hose Station Operators (Plug Man)

The AFFF hose station operators will serve as the plug man for each hose team. They are responsible for laying the AFFF hose out by unrolling it from the hose reel. When applicable, they will lay out fire hoses and use an AFFF inline eductor with 5-gallon AFFF cans. They shall establish and maintain communication with the AFFF proportioning station operator via the X50J phone circuit. They shall activate the AFFF hose or charge the fire station plug when directed by the hose team leader.

AFFF Proportioning Station Operator

This position is located in the AFFF stations below decks during flight operations or during crash and fire casualties. The AFFF station operator has the same responsibilities as the shipboard firefighting organization's AFFF station operator. The AFFF station operator maintains communications with the AFFF hose station operators via the X50J phone circuit. They shall monitor the AFFF levels and replenish the AFFF tank when necessary.

Rescue Personnel

Rescue personnel may sometime be referred as the hot suitmen because of the proximity suits they wear for personal protective gear. All rescue personnel shall be immediately available and dressed in their personal protective gear and breathing apparatus. Rescue personnel will position themselves near the scene leader for immediate dispatch once firefighting team has established a rescue path. They shall always work in pairs when conducting rescues and investigations. Rescue personnel responsibilities will include:

1. Rescue one incapacitated person at a time.
2. Be familiar with all aircrafts entry points.
3. Be familiar with all aircrafts shutdown procedures.

4 Be familiar with all aircrafts injection seat safeties.
5 Be familiar with all hazards associated with all aircrafts.
6 Conduct investigations for personnel casualties in surrounding areas.

Overhaul Personnel

The overhaul personnel perform overhaul procedures on aircrafts after the fire has been extinguished. Overhaul personnel are also known as hot suitmen because they must wear the proximity suit as well. Overhaul personnel shall be dressed in personal protective gear and breathing apparatus and positioned near the scene leader. Overhaul personnel will also work in pairs during overhaul evolutions. The senior overhaul person will carry the thermal imager camera (TIC) and the halligan tool or axe. The junior overhaul person will carry the CO_2 extinguisher to provide protection, extinguish residual fires on the aircraft and cool hot surface areas that are discover with the TIC. The rescue personnel may serve as overhaul personnel depending on the class of ship. Overhaul personnel shall be thoroughly familiar with the following:

1 Starting point to commence overhaul in relations to the liquid oxygen (LOX) converters and batteries access panels.
2 Location of batteries and LOX converters for each type of aircraft.
3 Disconnection of all aircraft batteries and LOX converters.
4 All hazards associated with each type of aircraft.
5 Operations of TIC.
6 Operations of portable extinguishers.
7 Operations of access tools.

Background Assistance Leader

The background assistance leader will assemble and organize the background assistance detail. This team is made up of the aviation fuel repair team personnel, embarked personnel and other aviation personnel not involved with flight deck crashes and fires. The background assistance leader shall be responsible for the following:

1 Ensures adequate flow of messengers to the scene leader.
2 Dispatches hose team reliefs for fatigued firefighters when required.
3 Assemble backup AFFF hose teams.
4 Ensure additional firefighting equipment and tools are available.
5 Remove aircrafts adjacent to the scene as necessary.
6 Provide medical personnel and stretcher bearers for injured personnel.
7 Provide firefighting teams if additional fires occur.
8 Provide replenishing hoses to the mobile firefighting vehicle for nursing operations.
9 Dispatches additional personnel for support as required by the scene leader.
10 Ensures that the fuel repair, electricians, and maintenance personnel are available.
11 Ensures external electrical power to aircraft involved is secured.
12 Ensures that elevators operators are in position.
13 Establish a SCBA refill station, when SCBA's are used.

Background Assistance Detail

The background assistance detail consists of the aviation fuel repair team, medical, electricians, maintenance and other embarked aviation personnel that are not directly involved with crash, rescue, and firefighting evolutions. Whenever the crash alarm or emergency flight quarters are announced, the detail will immediately assemble in the designated area and follow the directions of the background assistance leader. This detail is responsible for providing support to the crash and salvage team as well as personnel reliefs for fatigued personnel.

Messenger

The messenger will have a similar role as the shipboard firefighting organization's communicator. The messenger will position themselves directly behind the scene leader. The messenger is responsible for relaying information between the scene leader and appropriate control centers. Additional messengers may be used to ensure a continuous flow of communication is maintained from the scene to the appropriate control centers.

Medical Personnel

Medical personnel will be positioned with the background assistance detail and follow the directions of the background assistance leader upon sounding of the flight deck crash alarm or announcement of emergency flight quarters. The medical personnel will be assigned personnel from the hospital corpsman military occupation specialty and qualified stretcher bearers. They are needed to provide medical assistance as required. The will be responsible for the following actions:

1 Ensure designated area is prepared for the collection and treatment of injured personnel.
2 Treat all injured personnel from crashes and fires.
3 Provide direction to the stretcher bearers in providing CPR and first aid.
4 Transport injured personnel to medical treatment rooms, when required.

Weapons Personnel

During all flight operations with aircrafts involving ordnance, weapons personnel will be stationed in flight deck control or other designated area. Weapons personnel shall be personnel assigned from the military occupation specialties such as aviation ordnanceman, explosive ordnance disposal technician, or gunner's mate. During crashes and fires involving ordnances, weapons personnel shall be dispatched to the scene to provide technical assistance, ordnance cooling temperature checks, and ordnance disposal as required by the scene leader. The OHO shall maintain a status board that confirms type, quantity, and location of all weapons on the flight deck, hangar deck and aircraft. This information shall be provided to the scene leader, ACHO, and air boatswain.

Mobile Firefighting Vehicle (MFFV) Driver and Operators

The mobile firefighting vehicle (MFFV) is required during any aircraft operations such flights, maintenance turn ups, and fueling. The MFFV requires a qualified driver, turret operator, and one handline operator to provide immediate response and initial firefighting actions. The driver shall always approach the fire in the most effective direction to conduct firefighting operations. The turret operator directs the firefighting agent towards the fire when the vehicle is approaching or stationary. The handline operator will utilize the attached hose reel to apply agent onto the fire. The handline operator can also be used to connect replenishing hoses whenever directed.

Conflagration (CONFLAG) Station Operator

The CONFLAG station operator is utilized during operations on the hangar deck. They shall be thoroughly familiar with the operation of each AFFF sprinkling system zones or groups. The CONFLAG station operator shall activate appropriate zones of the sprinkling system under the supervision of the HDO or competent authority when multi-aircraft or spill fires are judged beyond the capability of the initial hose team. The CONFLAG station can prevent the fire from spreading throughout the hangar deck by closing hangar division doors to isolate the fire.

Training and Qualifications

Due to the unique characteristics of firefighting involving aircrafts, training is the essential link in developing an effective crash, salvage and rescue team. The unique characteristics involving aircraft fires bring forth challenges that may not be conquered without specified training and qualifications. Personnel assigned as crash, salvage, and rescue team members shall be qualified in various positions and attend formal training courses. The crash and salvage team members for each air-capable ship must attend the Shipboard Aircraft Firefighting Training Course. As a team, the crash and salvage team must attend the Aircraft Capable Ship Firefighting Team Training for qualifications during an 18 to 24-month cycle or whenever the team experiences a greater than 40-percent turnover. Both of these courses are held at various U.S. Navy training centers located throughout the U.S. and overseas. They shall receive additional in-depth training to include the following:

1. MFFV.
2. Personnel rescue procedures.
3. Hazardous ordnance cook-off times, weapons cooling, and jettison procedures.
4. Mobile crash handling equipment.
5. Aircraft entry (normal, manual, forced, and emergency).
6. Aircraft hoisting equipment.
7. Maintenance of crash handling equipment.
8. Crash dolly usage.
9. Boat and aircraft crane (when applicable).
10. Aircraft salvage procedures.
11. Aircraft jettison procedures.
12. Emergency flight and hangar deck repairs.
13. Aircraft familiarization.
14. SCBA.
15. Thermal imaging camera.
16. HAZMAT containment.

On-the-Job Training

The ACHO will ensure that all personnel assigned to positions during flight operations shall receive continuous training. The ACHO will also ensure embarked aircraft squadron personnel receive training as well. They shall receive on-the-job training from the subject matter experts and air department training team to include the following:

1. Organization and leadership of the crash, salvage, and rescue team.
2. Fire reporting procedures.
3. Communications.
4. First-aid and self-aid.
5. AFFF/saltwater station operation on flight and hangar decks including hangar deck sprinkler system.
6. Aircraft firefighting procedures.
7. Hazardous ordnance cooling and jettison procedures.
8. MFFVs (familiarization).
9. Catapult steam smothering.
10. Portable halon 1211, PKP, and CO2 extinguishers (operation and location).
11. Appropriate firefighting actions to perform until assistance arrives.
12. Basic handling of composite materials and hazardous materials produced after a crash or fire.

Hazardous Material Training

The hazardous materials used in aircraft structures and systems make firefighting very unique because special procedures must be used to protect the firefighters and extinguish the fire. All personnel assigned to the crash and salvage organization shall receive in-depth training specified in *NATOPS U.S. Navy Aircraft Firefighting and Rescue Manual, NAVAIR 00-80R-14* to ensure they are capable of handling hazardous materials produced after a crash or fire.

Fire-Involved Ordnance Training

Fires involving ordnance is one of the most dangerous hazards aboard Navy ships. There is a great risk that the ordnance will explode causing personnel death, injury, and further damage to the ship. Immediately cooling the ordnance is an important aspect of aircraft firefighting evolutions. All flight deck, hangar deck, and crash and salvage personnel should familiarize themselves with the various types of ordnance carried by embarked aircrafts. They should also be familiarized with ordnance's cook-off times specified in Appendix E of *NATOPS U.S. Navy Aircraft Firefighting and Rescue Manual, NAVAIR 00-80R-14*. Training and drills should be tailored to each type of ordnance that maybe used on aircrafts.

Aircraft Fire Drills

Aircraft firefighting drills shall be conducted with sufficient frequency to maintain the level of proficiency in the fundamentals of aircraft firefighting and salvage operations. All aircraft fire training drills should incorporate the following:

1 Class of fire, location, and aircraft damage.
2 Types of ordnance hazards.
3 Casualties (personnel and material).
4 When fire is under control.
5 When fire is extinguished.

ADTT shall implement drills that are challenging by incorporating "worse-case" scenarios. A few examples to make drills challenging is adding multiple ordnances, rupture AFFF or firefighting hoses, or MFFV's replenishing hose ruptures.

This chapter embraces the role that key members play in aviation fire protection. Because of the unique hazards associated with aircraft fires, the crash and salvage team must be thoroughly training in specialized firefighting, crash, and rescue evolutions. The key to an effective crash and salvage team begins with training, planning, leadership, and teamwork by both ship's company and embarked squadron personnel. Leadership and ADTT personnel should take advantage of every opportunity to drill and train aviation personnel in the ship's fixed and mobile firefighting equipment, aircraft configurations, fuel and weapons loads, and firefighting procedures specified within available technical manuals. The next chapter will cover the fundamentals of aviation fire protection.

CHAPTER 8

The Fundamentals of Aviation Fire Protection

Fire protection around crashed aircrafts requires highly specialized firefighting techniques. On fixed-wing aircrafts onboard nuclear-powered aircraft carrier (CVN), landing helicopter deck (LHD), landing helicopter assault (LHA), landing platform dock (LPD) class ships, and other air-capable ships, hazards to firefighters are connected with a wide scope of substances that range from fire-accelerating materials to non-combustibles. This scope is comprised of highly volatile aircraft fuels, ordnances, explosives, lubricants, fluids, propellants, and metals that will produce fires, explosions, dense smoke, toxic vapors, and hazardous debris. It is critical that firefighters acquire comprehensive skills and knowledge of the hazards that can affect the success of firefighting efforts, as well as protecting their overall health.

Aircraft Hazards

As mention in chapter 3, fires on board Navy vessels are classified as ALPHA (A), BRAVO (B), CHARLIE (C), or DELTA (D). The materials used in the operation and construction of aircrafts have the potential to become all four classes of fires when ignited. Firefighters require a basic knowledge of aircraft systems and each associated hazard that exists with aircraft systems. Significant systems within most aircrafts include the fuel system, hydraulic system, oxygen system, installed fire suppression system, and anti-icing system, and de-icing system. Many of the fluids that are routed throughout aircraft systems piping and tubing are hazardous. Recognizing the color schemes of the various identification labels used on piping, tubing, and conduits will enable firefighters to be more aware of the hazards present.

Fire-accelerating materials carried on aircrafts are major concerns to all aviation personnel involved with the operation and handling of aircrafts. Because fire acceleration materials are major concerns, firefighting and rescue efforts should begin immediately whenever there are aircraft crashes and fires. The common fire-accelerating materials associated with aircrafts include the following substances:

1. Jet fuels and aviation gasolines.
2. Oils, lubricants, and fluids.
3. Oxygen systems.
4. Class A combustibles.
5. Ordnance hazards.
6. Overheated batteries.
7. Flares, markers, and countermeasures.
8. Explosive suppressant foam (ESF).
9. Hydrazine.
10. Composite materials.
11. Radioactive materials.

Jet Fuels and Aviation Gasolines

Aircraft fuels and oils present the primary complications in aircraft firefighting evolutions. Aircraft fires may be cause from crashes and the fuel system may rupture from impact. Spilled fuel and hot crash debris often results in a fuel-fed inferno. The temperature produced by the burning of vaporized aircraft fuel and air is intense (approximately 1,500 °F). Oils may have greater heat retention properties which make them much more difficult to ignite than aircraft fuels. When heavier lubricating oils are combined with gasolines which will occur in aircraft crashes, the lubricating oils will become ignited because of the properties of the gasoline. The gasoline raises the temperature of the oil to the flash point, producing additional flammable vapors. The fuels that are commonly used in aircrafts are jet fuels and aviation gasolines (AVGAS).

The jet fuels used in aircrafts are JP-4, JP-5, and JP-8. JP-4 fuel is a blend of gasoline and kerosene. JP-4 fuel has a flashpoint of -10°F (-23°C) with a calculated flame spread rate between 700 to 800 feet per minute. JP-5 fuel is a kerosene grade with a flashpoint of 140°F (60°C) which is the lowest flashpoint considered safe for use aboard naval vessels. The rate of flame spread has been calculated to be in the order of 100 feet per minute. JP-8 fuel is a kerosene grade with a flashpoint of 100°F (38°C) and 100 feet per minute calculated rate of flame spread. Aviation gasoline (AVGAS) is fuel used in spark-ignited internal-combustion engines to propel aircrafts. AVGAS has a flashpoint is -50°F (-46°C). The rate of flame spread has been calculated to be between 700 and 800 feet per minute. These fuel fires can be extinguished using AFFF.

Oils, Lubricants, and Fluids

There are many oils and lubricants that are used in aircrafts. Oils and lubricants are less volatile and flammable than aircraft fuels and gasolines but still pose dangers to firefighters. The ignition of oils and lubricants are major concerns because of their location to hot engines and exhaust surfaces. When flammable hydraulic fluids are used to raise and lower landing gear, the risk of brake and tire fires will increase. The heat and dense smoke created by these fires can damage the aircraft, obscure vision, and create toxic fumes for personnel without breathing apparatus. These fires are Class B fires and AFFF can be used as the extinguishing agent.

The aircraft anti-icing systems used in aircrafts is usually a mixture of alcohol and glycerin. Anti-icing fluid is not as hazardous as other fluids and lubricants but must be considered during aircraft crashes and fires. This fluid is mostly made with alcohol and when it burns the flame is almost invisible creating a danger to firefighter. Care should be made when approaching the aircraft fire and water should be used to dilute the anti-icing fluid.

Oxygen Systems

Oxygen systems on aircrafts can present hazardous conditions to firefighters during firefighting evolutions. Liquid oxygen is a light blue liquid that flows like water and is extremely cold. It boils into gaseous oxygen at –297°F (–183°C) and has an expansion rate of approximately 860 to 1. Liquid oxygen is nonflammable but is a strong oxidizer that will actively support the combustion process of a fire. Liquid oxygen forms combustible and explosive mixtures when it comes in contact with flammable or combustible materials such as wood, cloth, paper, oil, or kerosene. Procedures for fighting fires involving liquid oxygen include cutting off the flow of oxygen and using large amounts of water at the seat of the fire.

Class A Combustibles

The aircraft's interior finish materials and electrical wiring within the aircraft are always potential fire hazards. The Aircraft's interior finish materials and wiring in the cockpit and cabin are made of various combustible materials. When interior finish materials and wiring are ignited, they produce toxic gases such as CO, HCl, and HCN; therefore, it is necessary that firefighting and rescue personnel who enter an aircraft during a firefighting and rescue evolutions be equipped with a breathing apparatus. Class A combustibles in aircraft fires are best extinguished with AFFF.

Ordnance Hazards

Naval aircrafts carry a wide variety of ordnance in support of their assigned missions and they present multiple hazards to firefighters. The first hazard is the cook-off of the warhead itself, but the firefighters must be aware of the propulsion systems of missiles, rockets, torpedoes, ammunition, and mines. Hypergolic fuels such as hydrazine are used as propellants in missiles and rockets while otto fuel and lithium are used in torpedoes and mines. These propellant demand special considerations when they are heated or ignited. Other hazards such as radioactive components in weapons also demand special consideration and precautions when encountered by crash, salvage, and rescue team members.

The aircraft ordnance exposed to a fire can cook off either during or after the fire is extinguished. The fire duration, the type, and location of the weapons will determine the reaction severity that may occur. Burning or heated ordnance may transition to detonation at any time and it is extremely important for personnel to take all safety precautions when approaching the fire.

Ordnance cooling should be immediately conducted whenever a crash or fire occurs involving an aircraft embarked with weapons. Ordnance cooling should be accomplished using a dispersed agent pattern that will provide coverage of agent over the weapon. The hose team should us extreme care by ensuring not to use a narrow stream pattern to cool the ordnance. Narrow stream pattern can dislodge hung ordnance or it can push or roll ordnance lying on the deck causing it to detonate.

The actual time for the cooling of weapons shall be exposure time of the weapon to the determined by the air boatswain based upon the actual fire. A general rule of thumb has been used for cooling weapons for a minimum of 15 minutes or until deemed safe by weapons personnel, but post assessment cooling for minimal exposures may be of a duration of less than 15 minutes, whereas severe heat exposures may require cooling in excess of 15 minutes. The use of a thermal imager can be used to find the surface temperature of the ordnance which can help determine the weapons cooling times.

Overheated Batteries

Non-lithium batteries, such as alkaline and nickel-cadmium, are found in various equipment and systems aboard aircrafts. These batteries present hazardous conditions to both aircraft and personnel when they become overheated when exposed to heat and fire. Overheated batteries may potentially release combustible gases and they may violently rupture or explode. The crash, salvage, and rescue team should immediately open battery compartments to check the battery for internal shortages or thermal runaways. Fires involving non-lithium batteries can be extinguished using halon 1211, PKP, AFFF or CO_2.

Lithium batteries are used in various aircraft and shipboard systems such as sonobuoys, weapons, and communication equipment. Lithium battery designs may vary in chemistry composition such as the following: lithium-thionyl chloride, lithium-sulfur dioxide, lithium-vanadium oxide, lithium-carbon monofluoride, lithium ion-cobalt oxide, and lithium-manganese dioxide. Because of the chemical composition of these batteries, highly flammable and toxic gases are released and when expose to heat and fire. Also, lithium batteries may potentially rupture and eject lethal fragments. The crash, salvage, and rescue team must approach with caution and immediately apply large amounts of water or AFFF to extinguish fires involving lithium batteries.

Flares, Markers, and Countermeasures

The Navy uses a variety of flares, markers, and countermeasures surrounding its aircrafts that produces emissions, charges, high intensity heat, high intensity light, infrared and high energy radio frequencies.

These devices may contain impulse charges, thermal batteries, lithium-ion batteries, pyrotechnic or pyrophoric material. These devices have unique characteristics when they are activated or ignited. These devices can produce temperatures up to 4000°F, bright lights and fumes that are extremely toxic. Crash, salvage and rescue teams responding to incidents involving these devices shall wear personal protective equipment.

The devices that contain pyrotechnic and pyrophoric materials will produce oxygen without depending on the oxygen in atmosphere. When pyrotechnic and pyrophoric materials are ignited agents such as PKP, AFFF and CO2 are ineffective in extinguishing the fire. The firefighting team will apply copious amounts of water or AFFF to the device until the device burns out. The water or AFFF is used to keep the area around the device cooled and prevent the device for igniting surrounding combustible material.

Explosion Suppression Foam (ESF)

Explosion suppression foam (ESF) is a flexible polyurethane foam material installed in certain aircraft fuel tanks and cells that provides protection against explosion. When ESF is ignited, it will produce toxic gases, intense heat, and dense smoke. The toxic gases that are produced from burning ESF are carbon dioxide, carbon monoxide, cyanides, and nitrous oxides. The crash, salvage and rescue team must ensure to wear breathing apparatus when conducting rescue and firefighting evolutions in the immediate vicinity or downwind of burning ESF.

Hydrazine

Hydrazine is a clear, oily, water-like liquid that smells like ammonia. Hydrazine is used as fuel for the F-16 aircraft's emergency power unit (EPU). Hydrazine fuel (H-70) is a blend of 70 percent hydrazine and 30 percent water. Hydrazine has a flashpoint of 126°F (52°C) and it will instantly ignite when exposed to heat, flame, or oxidizing agents.

Hydrazine vapors are much subtler to sparks, embers, or flames compared to its liquid form. Hydrazine is a strong reducing agent that becomes hypergolic with oxidizers. Spontaneous ignition may occur if it is absorbed in rags and other absorbing materials. Since hydrazine is highly flammable and may become hypergolic with exposure to oxidizers, its fires can best be extinguished by diluting with large amounts of water.

Composite Materials

Composite materials are now being used to replace heavy metal components used in the construction of aircrafts. Composite materials consisting of carbon-graphite fibers or boron fibers are reinforced to provide superior stiffness, high strength-to-weight ratio, and ease of fabrication. Carbon-graphite and boron fibers are released into the atmosphere once the epoxy that binds the fibers together begins to burn and melt. Whenever these small lightweight fibers are released into the atmosphere, they can be transported up to several miles by the smoke and wind. The epoxy used to hold the fibers together will ignite at approximate temperature of 752°F (400°C). A crash or explosion involving an aircraft will cause composite material fibers to become an immediate hazard. The crash, salvage and rescue team must immediately begin firefighting and containment operation for fires involving composite materials. Fires involving large amounts of composite materials can be extinguished using large quantities of AFFF. All personnel involved shall wear all protective garments and breathing apparatus.

Radioactive Materials

Aircrafts may be equipped with several components that may contain radioactive materials. These radioactive components do not present a significant hazard to personnel during normal operations of the aircraft. These radioactive sources may become a hazard during an aircraft crash, explosion, or fire in which radioactive material may disperse into the atmosphere. The crash, salvage and rescue team must ensure to adhere to all protective measures during firefighting, salvage or cleanup operations. The most significant hazard associated with mishaps involving radioactive material is the inhalation of radioactive particles. The firefighting clothing and breathing apparatus will protect all personnel involved with firefighting, salvage and cleanup operations.

Fire Protection Systems and Equipment

Because there are so many hazards associated with aircrafts, fire protection is a critical element during flight deck and hangar deck operations. All personnel involved shall be thoroughly trained in the operation of portable fire extinguishers, mobile firefighting vehicles, and fire protection systems. Maintenance, handling, and refueling personnel are the first line of defense for protecting naval assets whenever a fire or fuel spill occurs during aircraft evolutions. They must utilize all available fire protection systems and equipment to eliminate or contain the hazard until the crash, salvage and rescue team arrives on site. The available fire protection systems and equipment used on the flight deck and hangar deck are as follows:

1. AFFF system.
2. Portable fire extinguishers.
3. Visual aids.
4. Mobile firefighting vehicle (MFFV).
5. Personal protective equipment.
6. Firefighting and rescue tools.

AFFF Systems

The catastrophic incident that occurred on the *USS Forrestal* led to AFFF systems installations on all air-capable ships in the U.S. Navy. Both the AFFF proportioning and AFFF injections systems are used to supply the firefighting agent to all air-capable ship's flight decks and hangar decks. This system application has been installed to serve flight decks, deck edges, hangar deck overheads, AFFF hose reels, and AFFF hose outlets. Whenever flight operations are being conducted, AFFF stations that supply the flight deck have personnel positioned in the AFFF stations.

AFFF Proportioning Systems

Depending on the class of air-capable ship, the AFFF station will typically consist of an AFFF concentrate tank ranging between 50 to 600 gallons capacity, an injection pump, a proportioning unit, piping system, and electrical control system. Seawater and AFFF flow is controlled by hydraulically operated valves that are controlled by SOPVs that are activated by push buttons. The SOPVs are activated by electrical switches at pushbutton switches located at various control stations, AFFF hose reels, AFFF hose stations, and sprinkling groups. The most important component is the balanced pressure proportioner which mixes the AFFF concentrates with saltwater at a nominal appropriate foam concentration. The balanced pressure proportioner utilizes a 65 gpm pump and a balancing valve to mix AFFF concentrate and seawater to supply AFFF solution to the hose reels, hose outlets, flight deck nozzles, deck edge nozzles, and other sprinkling systems.

AFFF Injection Systems

The AFFF injection systems may be installed either as a single-speed injection or two-speed injection systems depending on the class of air-capable ship. The demand of AFFF will determine the output of the injection pumps. The single-speed pump will inject AFFF concentrate into the seawater supply line at rates of 12, 27 or 60 gpm. The two-speed pump will inject AFFF concentrate into the seawater supply line at rates of 27 or 65 gpm. The low-rate 12 and 27 gpm outputs serve AFFF hose reels, hose outlets, and small sprinkler system demands, whereas the high-rate 60 and 65 gpm outputs serve AFFF flight deck nozzles, deck edge nozzles, hangar deck overhead nozzles, and other large sprinkler systems.

AFFF Flight Deck and Deck Edge Sprinkling System

Air-capable ships' flight decks have an AFFF firefighting system consisting of flush deck and deck edge nozzles. The flush decks nozzles are strategically located in the flight deck surface and the deck edge nozzles are strategically located on the edges outlining the flight deck. Activation controls are located in various locations on the flight deck and secondary locations such as primary flight control, helicopter control station, navigation bridge, or damage control central all depending on the class of ship. AFFF solution for this system is supplied from the balance proportioner system or injection system. The system is designed to rapidly extinguish an aircraft fuel spill fire prior to heat buildup sufficient to initiate weapons cook-off conditions. It shall be activate immediately whenever any aircraft crash occur on the flight deck or when a fuel spill fire is discovered.

AFFF Flight Deck Weapons Staging Sprinkling System

The weapons staging area is protected by an AFFF sprinkling system consisting of deck edge nozzles that provide adequate AFFF coverage in the weapons staging area. The system is designed to rapidly extinguish an aviation fuel spill fire prior to enough heat buildup that may initiate weapons cook-off conditions. A single-speed injection pump will supply AFFF to this system. Controls to start and stop flow are located in primary flight control, navigation bridge, flight deck control, and on the fore and aft ends of the island structure on the flight deck.

AFFF Hangar Deck Sprinkling System

AFFF hangar deck sprinkling systems are installed in the overhead of the hangar bays of air-capable ships for embarked aircrafts. The configuration will vary for each class of air-capable ships. The sprinkler activation controls are typically located inside the hangar, outside the hangar, and inside helicopter control station on smaller air-capable ships such as CGs. DDGs, and LPDs. Large air-capable ships such as CVNs, LHAs, and LHDs sprinkler activation controls are located in the CONFLAG stations and throughout the hangar deck. Because the massive size of the hangar deck, the sprinkler system is divided into groups or zones on large air-capable ships. Each sprinkler group is activated individually in the CONFLAG station and in the vicinity of the related sprinkler group. Each group is supplied from two AFFF proportioning station: one located on the portside and the other located on the starboard side of the ship.

AFFF Hose Reel and Outlet Stations

AFFF hose reels and outlets stations are strategically installed to reach every area of flight decks and hangar decks during firefighting operations (Figure 8-1). The AFFF hose reels are usually colored red except the ones that serve the flight decks which are colored green. The station normally consists of one 1½-inch hose reel or one 2½-inch hose outlet with a vari-nozzle and multiple 1 ½-inch fire hoses. Each station has a pushbutton control located adjacent to the station. Air-capable ships with a helicopter landing platform usually installed AFFF hose reels on the port and starboard sides adjacent to the helicopter landing area. Air-capable ships that are equipped with helicopter hangars will normally have AFFF hose reels and outlets within the hangar area. Air-capable ships that have large flight decks will normally have AFFF hose reels and outlets located in the vicinity of island structure and in the catwalks. The placement of AFFF hose reels and outlets throughout the flight deck and hangar deck allow firefighting efforts to cover every area of the flight deck and hangar deck.

Figure 8-1. AFFF Hose Reel for Flight Decks and Hangar Decks

Portable Fire Extinguishers

Portable halon 1211, $CO_{2,}$ and PKP fire extinguishers are used on flight decks and hangar decks. There are specific requirements for the placement of these extinguishers throughout flight decks and hangar decks. Halon 1211 extinguishers are mounted in the MFFV and various locations in the vicinity of the flight deck. A CO_2 and PKP extinguisher shall be mounted at each AFFF hose reel and outlet station servicing the flight deck and hangar deck. Additional CO2 extinguishers fitted with an extended 5 or 7-foot applicator or wand may be provided for the flight deck on all air-capable ships. The extended applicator will be able to reach aircrafts engine compartments to extinguish fires during startups and shut downs.

Visual Aids

Visual aids have been added to flight decks and hangar decks as a quick reference for locating firefighting equipment. The deck edge coamings or wheel-stop coamings are marked with a 12-inch wide, color-coded stripe that goes up and over the wheel stop with 3-inch white lettering identifying locations of CO_2 extinguishers, PKP extinguishers, and AFFF hose reel stations. On those areas of the flight deck that does not have wheel stop coamings, the visual aid marking is placed on to the deck or on the bulkhead of the island structure. Hangar decks have visual aids placed high on the bulkhead throughout the hangar deck to easily identify firefighting equipment.

Mobile Firefighting Vehicle (MFFV)

The A/S32P-25 or P-25 for short is the MFFV used on large deck air-capable ships (Figure 8-2). The P-25 is a diesel-driven vehicle that is equipped with a turret, AFFF hand line, 55-gallon AFFF tank, 750-gallon water tank, proportioning pump, three portable halon 1211 extinguishers, and nursing connections. The high-pressure turret nozzle can discharge firefighting agent approximately 500 gpm. The discharge rate for the AFFF hand line and nozzle is approximately 60 gpm. The nursing connections are used to attach hoses to replenish the firefighting agent or connect directly into external sources such as the ship's AFFF system or firemain system.

The P-25 is utilized as a rapid response during flight operations, crashes, rescues and fires. The P-25 can simultaneously fight the fire, cool the cockpit area, and cool exposed ordnance until relieved by the crash, salvage, and rescue team. The MFFV driver shall immediately position the vehicle in a position that will provide the most effective method of controlling the fire, cooling ordnance, and protecting rescue personnel. Positioning the vehicle upwind of the fire is the most effective approach. There may be more than one P-25 used during flight deck causalities and the P-25 may take the place of a hose team. The P-25 can remain in place when the nursing hose is attached.

Figure 8-2. A/S32P-25 Mobile Firefighting Vehicle

Personal Protective Equipment

The greatest asset of the Navy is its personnel and protecting their health is a major priority. There are multitudes of hazards for personnel engaging in firefighting and rescue operations during aircraft crashes and fires. The proximity firefighting protective ensemble (PFFPE) is worn by the rescue personnel but can be worn by other team members when required (Figure 8-3). The SCBA is worn by rescue personnel or when required for other team members. Protective clothing is provided for all members of the crash, salvage, and rescue team. The protective clothing is different from the firefighter's ensemble. The crash, salvage, and rescue team members shall wear protective helmets, anti-flash hoods, firefighting gloves, steel-toed boots, inflatable life preservers, and fire retardant clothing in the form of shirts, trousers, or coveralls.

Figure 8-3. Proximity Firefighting Protective Ensemble (PFFPE).

The PFFPE, commonly referred to as the hot suit, is designed to protect firefighting and rescue personnel from extremely high temperatures that are produced by aircraft fires. The suit is made up of an aluminized material that will reflect heat and inner protective layers that will provide protection from the heat. The aluminized proximity fabric adopted for the use in the Navy mishap and rescue program. The hot suit comprised of the coat, trousers, helmet, gloves, boots, anti-flash hood, and summer aviator gloves. The SCBA can be worn with the hot suit when required.

The crash, salvage, and rescue team can easily be identified by their red protective clothing consisting of helmet, hood, life vest, and flight shirt. The protective helmet is unique because it not only provides head protection but also provides hearing and eye protection. The protective helmet is equipped with attached ear muffs and goggles. The anti-flash hoods are the same hoods worn with the firefighter's ensemble. The anti-flash hood shall be worn over the wearer's nose to protect the nose from being burned and inhaling composite fibers. The firefighting gloves are the same gloves worn with the firefighter's ensemble. Steel-toed boots approved for flight deck usage are worn to protect the feet. Inflatable life preservers are worn by the crash and salvage team members in case personnel fall into the ocean. The life preserver will automatically inflate once it is introduced to saltwater. Also, the life preserver is equipped with a dye marker and strobe light to easily identify personnel in the water during rescue. The flight shirts, trousers, or coveralls are made of flame resistant material. The flight shirts are long sleeve and worn with the trousers or overtop of the coveralls.

Firefighting and Rescue Tools

There are tools that are available for the crash, salvage, and rescue team to aid them during firefighting and rescue operations. Tools such as hold control devices will aid with firefighting efforts. There are tools to help with gaining access into aircrafts to rescue personnel and investigate for hidden fires in aircraft compartments. The classification of air-capable ships determines the type of tools required for the tool inventory.

Hose control devices are used on CVN, LHA, and LHD platforms to assist with cooling ordnance. It is designed to be securely fastened to the deck pad eye. A navy standard firefighting nozzle and hose is connected to this device and directs the flow of AFFF or water towards any exposed or at-risk ordnance during a fire on the flight deck or hangar deck. This device is used to minimize the number of personnel expose to a cook-off hazard of ordnance. After the device is attached and the agent is flowing on the ordnance, personnel can move away from the potential explosion hazard. During windy conditions, one person must stay with the device to adjust the stream of the agent to compensate for the winds blowing the stream off the target. The hose control device is available in 1 ½ and 2 ½-inch sizes.

A crash locker containing the firefighting and rescue tools and equipment shall be maintained for use in emergencies. All equipment shall be inspected daily prior to commencement of flight operations. Typically, the crash and salvage tool roll is most common across all the air-capable ship platforms. The tool roll shall contain a minimum of one each of the tools listed below:

1. Pliers locking (vice grip) 8-inch.
2. Pliers, side cutting 8-inch.
3. Pliers 10-inch (slip-joint).
4. Pliers 6-inch (slip-joint).
5. Pliers, long nose (needle nose).
6. Punch, drift (8-inch).
7. Saw, hack frame with 6 spare blades.
8. Screwdriver, Phillips #2.
9. Screwdriver, Phillips #3.
10. Screwdriver, flat tip 3/16-inch tip, 6 inches long.
11. Screwdriver, flat tip 5/16-inch tip, 10 inches long.
12. Wrench, 8-inch adjustable.
13. Halligan tool.
14. Strap cutter (rescue knife).

This chapter describes the unique hazards associated with aircraft crashes and fires. Fire protection is critical in safeguarding aircrafts, ships and personnel. The crash, salvage, and rescue team must be thoroughly familiar with all the hazards involved with aircrafts and its weapons. They must be able to maintain and operate all available firefighting and rescue equipment. The AFFF system supplies agent to the flight deck and hangar deck by use of sprinklers and hoses. The MFFV provides a rapid means for the crash, salvage and rescue team to rapidly fight a fire and cool ordnance. The PFFPE and rescue tools enable the rescue team to safely access aircrafts and rescue personnel during firefighting operations. The next chapter will cover the firefighting procedures use for fires involving aircrafts.

CHAPTER 9
Aircraft Firefighting Tactics

The flight deck fire onboard the *USS Forrestal* change firefighting systems and procedures utilized on air-capable ships. Firefighting procedures may be well defined but there are no identical situations during fires on the flight deck and hangar deck. Successful firefighting operations will always depend on training, planning, leadership, and teamwork by both ship's personnel and embarked air wing personnel. Crash, salvage, and rescue team and ADTT members should take advantage of every opportunity to conduct training and drills. The training should be tailored towards their ship's fire protection systems, firefighting equipment, aircraft configurations, fuel, and weapons loads. Drills should focus on using specified firefighting tactics and procedures to enable the crash, salvage, and rescue team to respond to the worse-case scenarios.

Firefighting Tactics

When fire occurs on the flight or hangar deck, the casualty shall be passed over the 1 MC announcing system. If the fire is a result of an aircraft crash, the flight deck crash alarm shall be sounded to notify flight deck personnel of an actual on-deck aircraft mishap. Designated personnel shall report to the AFFF proportioning stations and establish communications with primary flight deck control station or helicopter control station. The crash, salvage, and rescue team must immediately respond to rescue personnel and extinguish the fire. Immediate response will save lives and prevent fires from spreading to the below decks causing damage to the ship's interior.

The OSL must ensure that the hose teams are following specified firefighting procedures for each scenario. Firefighting tactics used are solely dependent upon the following varying factors.

1 Flight deck obstacles.
2 Wind direction.
3 Type of aircraft.
4 Crew stations and passenger locations within the aircraft.
5 Fire location on aircraft or the degree of fire involvement.
6 Presence of ordnance.
7 Presence of hazardous cargo.

The OSL has a checklist that they must be thoroughly familiar with during firefighting operations. The OSL checklist consists of commands to the hose teams and verbal reports that must be made to primary flight deck control or helicopter control. The OSL checklist shall include the following steps when there is a fire on the flight deck or hangar deck:

1 The OSL shall ensure that all firefighting equipment and personnel are available. The MFFV will be used on CVNs, LHAs, and LHDs. Four hose teams with team leaders must be available for CVNs, LHAs, and LHDs. Three hose teams with team leaders must be available for other air-capable ships. The hose teams will ensure that the

AFFF firefighting hoses are laid out. The rescue personnel must be available and dressed in the PFFPE. Once all firefighting equipment and personnel are in place, the OSL will make a "manned and ready" report to primary flight deck control or helicopter control.

2 The OSL will give the order "nozzles on, move in!" The OSL may order the MFFV to back out in order for the hose teams to hear the OSL commands. A hose teams may replace the MFFV once it backs out. The OSL will report "hose teams moving in" to primary flight deck control or helicopter control.

3 The OSL will order the designated hose team to commence weapons cooling. The OSL will report "weapons cooling in progress" to primary flight deck control or helicopter control.

4 The OSL will order hose teams to "establish rescue path!" Once rescue path is establish, the OSL will order the rescue team to conduct rescue. The OSL will report "rescue in progress" to primary flight deck control or helicopter control.

5 Once all personnel are removed from aircraft by the rescue team, the OSL will report "rescue complete" and " (number of __) personnel rescued" to primary flight deck control or helicopter control.

6 The OSL will order "hose teams, move in!" The hose teams will combat the fire until the fire is out.

7 Once the hose team leaders make the "fires out" report, the OSL will report "fires out" to primary flight deck control or helicopter control.

8 The OSL will order the designated weapons cooling hose team to continue cooling weapons for 15 minutes.

9 The OSL will make the order "nozzles off, back out!" to the hose teams.

10 The OSL will request weapons personnel to report to the scene to check the status of the ordnance.

11 Once weapons personnel on the scene, they will determine if the weapon is safe or cooling shall be continued. The OSL will report "weapon safe" or "continue cooling" to primary flight deck control or helicopter control. Weapons personnel will determine to download and chain down ordnance or jettison the ordnance.

12 The OSL will order the overhaul team to commence overhaul and battery disconnect procedures. The OSL will report "overhaul in progress" to primary flight deck control or helicopter control.

13 Once the overhaul team disconnects all the aircraft batteries, the OSL will report "batteries disconnected" to primary flight deck control or helicopter control.

14 Once overhaul procedures are completed, the OSL will report "overhaul complete" to primary flight deck control or helicopter control.

15 The OSL will order a designate hose team to set the reflash watch. Once the reflash watch is set, the OSL will report "reflash watch set by hose team number (__)" to primary flight deck control or helicopter control.

16 The OSL will order a foreign object of destruction (FOD) walk down of the flight deck or hangar deck. The personnel will be looking for objects and debris that will cause damage to aircraft engines.

17 Once all crash debris and FOD is removed and the affect aircrafts has been cleared, the OSL will report "ready deck" to primary flight deck control or helicopter control.

The firefighting tactics and procedures shall be deployed to extinguish the following type of fire scenarios:

1 Aircraft brake and wheel assembly fires.
2 Aircraft tailpipe fires.
3 Aircraft accessory, compressor, and engine compartment fires.
4 Aircraft internal engine fires.
5 Aircraft engine wet start fires.
6 Aircraft electrical and electronic fires.
7 Hangar deck fires.
8 Fuel station fires.
9 Aircraft debris and running fuel fires.
10 Helicopter fires.
11 Aircraft jettison.

12 Weapons staging area fires.
13 Multiple aircraft conflagration.

Aircraft Brake and Wheel Assembly Fires

During aircraft landings, the landing gear is an item of significant concern because of the extreme braking required on shorter runways. The weight of the aircraft and extreme braking can potentially cause brakes and wheels to become overheated and ignite in a fire. Overheated aircraft wheels and tires present a potential explosion hazard because of built-up air pressure in the tires. The explosion hazard is greatly increased when fire is present. Pieces of the wheel and tire are projected into the air when wheels and tires explode. The area should be immediately evacuated to avoid endangering the personnel.

Overheated brake and wheel assemblies must be cooled down. The recommended cooling procedure is to park the aircraft in an isolated area and allow the assemblies to cool in the surrounding air. Using water as a cooling agent is not recommended unless absolutely necessary whenever personnel has to be near the overheated assembly. Propeller-driven aircrafts can use the propeller to provide an ample airflow to cool overheated assemblies. Most major jet, propeller-driven, and turboprop aircraft now have fusible plugs incorporated in the wheel rims. These fusible plugs are designed to automatically deflate the tires and the reduced pressure on the wheel eliminates the possibility of explosion.

When responding to a wheel fire or hot brakes the crash, salvage, and rescue team shall approach the wheel with extreme caution. Personnel should approach the wheel and brake assembly from the forward or aft directions. Personnel should never approach the wheel and brake assembly from the side directions in line with the wheel axle. Peak temperatures may not be reached until 15 to 20 minutes after the aircraft has come to a complete stop. Materials that may contribute to wheel assembly fires are grease, bearing lubricants, hydraulic fluid, and tire rubber. Since heat is transferred from the brake to the wheel, agent application should be concentrated on the brake area. The main objective is to prevent the fire from spreading upward into wheel wells, wing, and fuselage areas. The following extinguishing methods shall be applied to wheel assembly fires:

1 Wheel grease and bearing lubricant fires can be identified by long flames around the wheel brake and axle assembly. These fires are usually small and should be extinguished quickly using halon 1211, PKP or water fog.
2 Rubber from the tires may ignite at temperatures from 500°F (260°C) to 600°F (316°C) and can develop into an extremely hot and destructive fire. Halon 1211, PKP, or water fog should be used immediately to extinguish the fire. The fire may reignite once the rubber sustains its auto-ignition temperature or the rubber is abraded and the fire is deep seated.
3 A broken hydraulic line may result in the flowing of petroleum-based fluids onto a damaged or hot wheel assembly. Upon ignition, flowing hydraulic fluid will accelerate a fire causing rapid fire growth and excessive damage to the aircraft if it is not extinguished rapidly. An application of water fog in intermittent bursts (5 to 10 seconds) every 30 seconds should be used as the extinguishing method.

The crash, salvage, and rescue team must take cautions around fires involving the wheel and brake assembly. The following precautions must be taken when conducting firefighting operations:

1 Rapid cooling may cause an explosive failure of a wheel assembly.
2 Effectiveness of halon 1211 may be severely affected if conditions are such that sufficient agent concentrations cannot be maintained on the fire source.
3 Halon 1211 may extinguish hydraulic fluid fires but lacks adequate cooling effect resulting in the fire reigniting.
4 Some Navy aircraft and most commercial aircraft incorporate hydraulic systems that contain hydraulic fluid by the trade name of Skydrol. When expose to fire, Skydrol will not ignite but will decompose and produce vapors that will cause severe irritation to

the eyes and respiratory tract. Firefighters shall wear SCBA when fighting fires involving Skydrol.

5 Aircrafts with beryllium brakes may produce toxic gases that cause irritation to the eyes and respiratory tract when exposed to fire. Firefighters shall wear SCBA when fighting fires involving beryllium brakes.

Aircraft Tailpipe Fires

During shutdown procedures of an aircraft, there is a potential for a fire in the aircraft's tailpipe. Whenever this occurs, authorized personnel will start the aircraft engine to attempt extinguishing through exhaust pressures. If this procedure does not extinguish the fire, the crash, salvage, and rescue team member will performed the following extinguishing procedures:

1 Direct halon 1211 or CO_2 fire-extinguishing agents into the aircraft's tailpipe.

2 If fire is not extinguished by the above method, direct the stream of extinguisher agent into the intake duct. Personnel should not stand directly in front of intake duct or directly in back of tailpipe.

3 AFFF can be used to extinguish the tailpipe fire if the above agents do not extinguish the fire or if the fire has spreaded to other areas of the aircraft. PKP can be used if AFFF is not readily available.

Aircraft Accessory, Compressor, and Engine Compartment Fires

Fires may occur in the accessory section, compressor section or engine compartment of jet aircrafts whenever fuel or hydraulic is introduced into the area between the equipment housing or between the engine and fuselage. Fires can ignite in these areas because of the heat that is generated by the engine or auxiliary power unit. Halon 1211 or CO_2

is the preferred extinguishing agent used on these fires. AFFF can be used to extinguish these fires when these fires cannot be extinguished with halon 1211 or CO_2.

The following procedures are conducted to extinguish fires in the aircraft accessory section:

4 Halon 1211 or CO_2 may be introduced into the engine accessory section area through the access doors located on the aircraft engine cowling.

5 When the fire is under control, one firefighter in a PFFPE will open the engine cowling with a screw driver, if required. An AFFF handline should be used to provide fire protection to the firefighter.

The following procedures are conducted to extinguish fires in the aircraft's compressor section engine core:

1. Halon 1211 or CO_2 may be introduced into the engine intake, exhaust, or accessory section.
2. When the fire is under control, one firefighter in a PFFPE will open the engine cowling with a screwdriver, if required. An AFFF handline should be used to provide fire protection to the firefighter.
3. When the engine cowling is open, apply AFFF to both sides of the engine casing to complete extinguishing and provide additional cooling.

Internal Engine Fires

Internal engine fires usually occur when residual fuel is dumped into the engine during shutdown. When qualified personnel are immediately available to start the aircraft's engine, these fires may be controlled by a procedure known as wind-milling the engine. Halon 1211 or CO_2 extinguishing agent must be directed into the engine if qualified personnel are not available or if wind-milling the engine fails. If halon 1211 or CO_2 agents are not effective or not available, AFFF should be used and PKP should be used if AFFF is not immediately available.

The following procedure is used to extinguish internal engine fires:

1 Application of halon 1211 or CO_2 must be accomplished at a distance.
2 Agent should be applied through the air intake, but can be applied through the tailpipe depending on the location of the fire in the engine and wind direction.
3 Apply agent by swirling the extinguisher nozzle to fill the entire engine space with firefighting agent.

Aircraft Engine Wet Start Fires

Aircraft engine wet start fires are caused when accumulated residual fuel ignites within the engine or tail assembly area. Fire, smoke, and heat may exit from the intake or exhaust areas of the aircraft depending upon the amount of ignited fuel and the direction of the wind. This type of fire occurs during initial engine startup of the aircraft. Wet start fires can be extinguished when authorized personnel increase the engine rpm to blow the fire out. Halon 1211 or CO_2 agent is used if this procedure is not successful. The following procedures are used to extinguish this type of fire:

1 Approach affective aircraft from the windward side with CO_2 or halon 1211 fire extinguisher.
2 Direct CO_2 or halon 1211 agent into the intake.
3 Attacking the tailpipe may be attempted if engine is not turning.
4 AFFF or PKP can be used if CO_2 or halon 1211 is not available.

Aircraft Electrical and Electronic Fires

Electrical power must be immediately secured whenever there is a fire in the aircraft's electrical and electronic systems. Halon 1211 and CO_2 agents are the primary extinguishing methods for combatting Class C fires. These agents do not have any adverse effect on electrical or electronic components. If fire cannot be extinguished using the primary agents available, PKP may be used to prevent the risk of fire spreading and causing further damage to the aircraft. The following precautions must be followed when fighting Class C fires on aircrafts:

1 Halon 1211 may be used to inert a small electronics compartment provided firefighters do not enter the compartment or do so with self-contained breathing apparatus.
2 The firefight must not use CO_2 to inert the atmosphere in an electronics compartment as it may produce a spark.

Hangar Deck Fires

The hangar deck is used to store aircrafts, ordnance, and equipment. Hangar deck fires are very unique because it is located inside the ship. There is a great risk of fires spreading to the ship's lower decks if firefighting operations are not started immediately to control fires in the hangar deck. The crash, salvage, and rescue team may require assistance from the Flying Squad to fight fires in the hanger deck. The following additional procedures for aircraft fires on the hangar deck shall be followed:

1 Return elevators to the flight deck level.
2 Close divisional doors immediately.
3 Close all doors and hatches from the hangar to the interior of the ship.
4 Close all weapons elevator doors and hatches.
5 Open elevator doors to provide ventilation to help remove heat and smoke.
6 Leave all hangar deck lights on.
7 Initial response personnel without SCBA shall be immediately relieved by the crash, salvage, and rescue team or Flying Squad members with SCBA.
8 The OSL will access the fire and determined the method of attack. The methods to use will either be direct attack at the seat of the fire, fog attack to control the fire, or direct attack from the access.

9 Background assistance should be established in an adjacent unaffected hangar bay.

10 Cooling teams shall be posted on opposite side of divisional doors of affected bay to monitor temperature and cool as necessary.

11 Activate appropriate zones of the hangar bay AFFF sprinkler system for any multi-aircraft fire or when a fire is judged to be beyond the capability of the hose teams or if determined necessary to minimize collateral damage.

Fuel Station Fires

Fuel station fires bring forth unique challenges when combating internal fuel station fires. The challenges are the layout and obstacles firefighters face when responding to the fuel station fires. Personnel shall exercise extreme caution when conducting firefighting efforts in proximity to the deck edge and catwalks. Entry into the catwalk shall not be attempted by ladders that are in close proximity to the fire. Ladders downwind of the fire should not be used. Firefighting tactics used for aircraft fires do not apply to fuel station fires. Firefighters must ensure to keep a low profile to avoid heat and heavy black smoke produce by fuel fires. Firefighters should conduct the following procedures and tactics for fires that affect fuel stations serving the flight deck:

1 Upon notification of fire or smoke in a fuel station, JP-5 pumping operations shall be immediately stopped throughout the ship. Damage control central will dispatch Flying Squad personnel to establish internal fire boundaries. Fuel repair team personnel shall mechanically and electrically isolate the fuel station and notify flight deck control.

2 The MFFV should make an upwind approach and utilize the turret and AFFF handline to affect initial attack of the fire and protect nearby aircraft.

3 All attempts should be made by crash, salvage and rescue team personnel to insert an AFFF hose with the nozzle set for wide angle fog

into the station porthole. If heat and smoke becomes too intense, the hose and nozzle can be pushed further through the porthole into the station allowing hose team personnel to fall back and observe the hose from a safe distance until properly attired crash, salvage, and rescue team personnel arrive. Initial response should be based on the one-person activation and operation concept for AFFF hose reels.

4 As additional hose teams form up at the scene they should be deployed in a manner to prevent fire from being able to extend beyond the station, impinging on nearby aircraft and to keep the deck cool. Intermittent applications of AFFF can be used for deck cooling.
5 The first two crash, salvage, and rescue team personnel dressed in a PFFPE with SCBAs shall take over the firefighting efforts at the porthole and adjust the nozzle pattern so a blanket of AFFF is applied throughout the fuel station.
6 The second two crash, salvage, and rescue team personnel dressed in a PFFPE with SCBAs shall take over the nearest AFFF hose and advance toward the fueling station entrance to attack the fire from the doorway.
7 At the discretion of the scene leader and air boatswain the firefighting team at the porthole will secure firefighting operations to help improve firefighting conditions from the door and allow for ventilation.
8 When the fire is out, the second hose team will remain at the door to provide reflash protection for the overhaul team to conduct overhaul procedures.

Firefighters should conduct the following procedures and tactics for fires that affect fuel stations serving the hangar deck:

1 Upon notification of fire or smoke in a fuel station, JP-5 pumping operations shall be immediately stopped throughout the ship. Damage control central will dispatch Flying Squad personnel to

establish internal fire boundaries. Fuel repair team personnel shall mechanically and electrically isolate the fuel station and notify flight deck control.

2 Hangar deck personnel in the area shall immediately activate the closest AFFF hose reels and respond in order to prevent fire from extending beyond the station and impinging on nearby aircraft by directing agent at the door from a safe distance.

3 Activate fuel station AFFF sprinkling (as equipped/required).

4 Divisional doors shall be closed immediately. Elevator doors shall be configured to vent heat and smoke out of the hangar.

5 Remaining hangar deck personnel shall evacuate the hangar bay. They shall immediately put on SCBAs, flash hoods, and gloves and relieve the non-protected personnel on the AFFF hose lines.

6 As additional firefighters dressed in SCBAs, flash hoods, and gloves arrive at the scene, they shall relieve the initial response personnel. They will discharge AFFF into the fueling station to prevent the fire from spreading and to protect nearby aircraft. Intermittent applications of AFFF should be used for bulkhead cooling.

7 Refuel repair team and crash, salvage and rescue teams dressed in complete PFFPE with SCBAS shall arrive to the scene and report directly to the OSL for tasking. The scene shall be turned over to the ship's Flying Squad while the hangar deck personnel provide assistance and support as directed by the Flying Squad OSL.

8 The Flying Squad shall assume the responsibilities for firefighting, overhaul, de-smoking, de-watering, and atmosphere testing.

Aircraft Debris and Running Fuel Fires

Aircraft debris and running fuel fires normally occur during catastrophic aircraft crashes where the aircraft's fuselage and wing fuel cells have been torn open. There is a great possibility that multiple aircrafts may be involved. These situations can lead to aerated fuel to flow deep down into aircraft debris which will often shield the fire from direct attack with

AFFF by crash, salvage, and rescue hose teams. These aircraft debris pile and running fuel type fires can easily become self-generating if fuel is fed from ruptured tanks. This will generate more fire causing additional fuel cells to be degraded and opened which will result in a growing fire with high flames and very intense heat. Getting this type of fire under control will require a highly organized effort among the OSL and hose teams along with effective use of the ship's fixed AFFF flush deck systems. The following tactics shall be applied:

1 Activate AFFF flush deck zones at the scene of the fire as well as other AFFF zones that is upwind of the fire.
2 Using AFFF agent, the crash, salvage, and rescue hose teams and MFFV should approach the fire with the wind at their backs ensuring to extinguish outlying pool and residual fires. As the ship turns to provide optimal winds, the hose teams and MFFV shall respond accordingly.
3 The hose teams should attack the fire from a 45° angle.

Once the fire is contained, the OSL will assemble the hose teams close together to conduct a coordinated attack on the weakest and most advantageous point of approach to the debris fire. The hose teams will approach the debris fire to provide both high-point cooling and low-point AFFF firefighting coverage. This is best accomplished with two nozzles on a wide fog pattern to block the radiant heat and two nozzles on a narrow fog pattern to knock down and extinguish the fire. Their approach should be dedicated to getting as close to the seat of the running fuel fires as possible. This is accomplished by a coordinated advance on the fire combined with a methodical sweeping of the fire area.

Once the hose teams have moved close to the fire and their firefighting efforts have diminished the intense heat and flames, a hose team member with a portable halon 1211 or PKP extinguisher will enter between the two low-point AFFF hose teams. The hose team member will use the portable extinguisher to expel halon 1211 or PKP agent onto the seat of the fire. This procedure may be repeated until the fire is extinguished. Hose teams shall continue cooling all fire-related debris to prevent reflash from the deep-seated embers or super-heated metal parts. This will continue until directed to stop by the OSL or competent authority.

Helicopter Fires

Whenever there is a helicopter crash or fire, the flight deck crash alarm will sound over the ship's announcing system. The crash, salvage, and rescue team shall take cover until the flight deck crash alarm stops. The flight deck crash alarm will keep going until the helicopter rotor blades stop rotating. There is a great possibility that the flight deck will be engulfed in flames because of the helicopter fuel spillage. The following procedures shall be conducted to extinguish the helicopter fire:

1 The HCO or primary flight control personnel will activate the appropriate flight deck AFFF sprinkling system zones.

2 The HCO or primary flight control will make a request for the ship to rotate to get favorable wind conditions.

3 The OSL shall assess the severity of the helicopter fire and direct the hose teams to approach the fire as rapidly as possible with the wind to their backs. If there are and ordnance involved, the designated hose team shall immediately commence ordnance cooling operations.

4 The hose team leaders shall order the nozzleman to test for AFFF agent and order the hose team to commence firefighting operations. The hose teams will separate to attack the fire on each side of the helicopter.

5 The hose teams shall create a rescue path by fighting the fire to a position where there is enough clearance for the rescue personnel to approach the aircraft. The hose teams will maintain their positions on the fire while rescue operations are in progress.

6 When directed by the OSL, the rescue persons (hot suitmen) shall work as a team to evacuate the aircrew and passengers. Each team of rescue persons shall maintain a two-man buddy system

throughout the rescue. In multiple casualty rescues, the team shall concentrate on evacuating one person at a time. During removal of the occupants, the OSL shall continue to direct extinguishment of the fire. The hose teams shall be careful not to get the rescue personnel wet. If rescue personnel become wet during entry, the hose teams shall continuously cool rescue personnel to prevent scalding.

7 Once rescue has been completed, the OSL will order the hose teams to continue to fight the fire until the fire is out.

8 Once hose team leaders report the fire out on each side of the helicopter. The OSL will order the hose teams to back out.

9 Once each hose team has backed out, the OSL will order one of the hose teams to set the reflash watch. The designated hose team will move to the center of the helicopter. This will allow them to quickly advance on the fire if a reflash occurs.

10 The OSL will order the overhaul personnel to commence overhaul of the helicopter. The hot suitmen will work in pairs to disconnect the helicopter battery and check for thermal runaway. If thermal runaway occurs, the hose team that was designated as the reflash watch will extinguish the battery fire.

11 The ordnance cooling team will continue to cool the ordnance for 15 minutes after the fire is out or until weapons personnel determine that the ordnance is safe.

12 Once overhaul is completed, the air boss shall determine to keep the helicopter or to salvage and jettison it.

13 The OSL will make a recommendation of the time it will take for the flight deck will be ready for flight operations. The OSL will make this report to the HCO or primary flight deck control.

Aircraft Jettison

Aircraft jettison or removal may be required if an aircraft fire becomes uncontrollable and endangers the ship. The commanding officer is the only authority to approve an aircraft jettison. The procedures used will determine on the class of ship and type of aircraft. The procedures for jettison aircrafts are as follows:

1 The ship can use high-speed, full-rudder turns to create centrifugal forces of sufficient magnitude to roll the helicopter over the side of the ship.
2 Cranes or forklifts can be utilized.
3 The ship's capstan can be utilized.

Weapons Staging Area Fires

Crash, salvage and rescue team shall respond to a fuel spill or fire on the flight deck's weapon staging are when ordnance is stowed. The hose team shall use AFFF agent to conduct rapid fire extinguishment or provide weapons cooling protection. The OSL should evaluate the threat and recommend activation of bomb farm AFFF sprinkling system and adjacent AFFF flush-deck sprinkler zones as necessary. The hose team shall conduct the following procedures:

1 As hose teams arrive on the scene, they shall knock down fire and smoke to enable identification of fire-exposed ordnance.
2 Once fire-exposed ordnance is identified, the hose team shall lock their agent on that particular ordnance.
3 Hose team leaders shall ensure minimum manning of hose team personnel and ensure hose team personnel maintain a low profile as possible should a deflagration or explosion occur.
4 Continue cooling of fire-exposed ordnance for 15 minutes after all residual fire and smoke has ceased or until weapons personnel have determined that ordnance has reached safe ambient temperatures and all ordnance inventoried.

Multiple Aircraft Conflagration

The magnitude of multiple aircraft conflagration is the loss of lives, multiple personnel injuries, multiple fires, severe damage to the ship's structure. In the event this occurs on the flight or hangar deck, procedures for life safety, fire suppression and damage control must commence immediately. This event may cause the ship to set general quarters. The following procedures will be employed to combat a multiple aircraft conflagration:

1 The initial response to the scene shall include MFFVs and crash, salvage, and rescue hose teams.

2 All available stretcher bearers shall supply stretchers and first-aid kits. The medical department will establish battle dressing stations and conduct mass casualty procedures, as required.

3 The background and assistance team will continuously supply and augment portable extinguishers and firefighting equipment to the scene as required by the OSL.

4 OSL will determine to conduct nursing procedures for the MFFVs.

5 The air officer will activate appropriate zones of the flight deck AFFF sprinkling system for any multi-aircraft fire or when a spill fire is judged to be beyond the capability of the initial hose team and MFFV.

6 The ACHO establishes the area for collection and disposition of personnel casualties. The ACHO will designate elevators to move personnel casualties. Also the ACHO will ensure unaffected aircrafts are move to safety.

7 The HDO shall man the applicable CONFLAG control station. The HDO shall order the CONFLAG station operator to activate the appropriate zones of the hangar deck AFFF sprinkling system for any multi-aircraft fire or when a spill fire is judged to be beyond the capability of the initial hose team.

8 Assistance Flying Squad or DCRS may be needed, as determined by the air officer.

9 Firefighting operations will continue until fires are extinguished.

The primary mission of the crash, salvage, and rescue team is to save lives, extinguish aircraft fires, and protect the flight deck and hangar deck from damage caused by fires. The numerous hazards associated with aircrafts are very unique and no fire situations will be identical. There are many fire scenarios that the crash, salvage, and rescue team must be prepared for by being thoroughly familiar with all the fire-fighting procedures and tactics used on the flight and hanger decks. A successful crash, salvage, and rescue team solely depends on leadership, training, planning, and teamwork by both ship's company and embarked air wing personnel. Once the flight deck crash alarms sound all personnel involved must immediately take actions to rescue personnel and extinguish fires to ensure that the ship stay on course to meet its mission.

References

1 Surface Ship Survivability, NTTP 3-20.31, Office of the Chief of Naval Operations, Washington, DC, 2012.

2 Naval Air Training and Operating Procedures Standardization (NATOPS) U.S. Navy Aircraft Firefighting and Rescue Manual, Naval Air Systems Command, Patuxent River, Maryland, April 2012.

3 Naval Ships' Technical Manual (NSTM), Volume 3, Chapter 074, Gas Free Engineering, Naval Sea Systems Command, Washington, District of Columbia, April 2011.

4 Naval Ships' Technical Manual (NSTM), Volume 1, Chapter 555, Surface Ship Fire Fighting, Naval Sea Systems Command, Washington, District of Columbia, September 2010.

5 Naval Ships' Technical Manual (NSTM), Volume 2, Chapter 555, Submarine Fire Fighting, Naval Sea Systems Command, Washington, District of Columbia, September 2007.

6 Damage Controlman, Naval education Training Manual 14057, Naval Education and Training Program Development Center, Pensacola, Florida, 2003.

Appendix I – Glossary

AIR DEPARTMENT TRAINING TEAM (ADTT): The air department training team (ADTT) consists of subject matter experts with strong backgrounds in crash, salvage, and rescue evolutions. This team will be responsible for training the crash and salvage team through on-the-job training and drills. Air-capable ships that do not have an air department will have a designated aviation training team (ATT).

AFFF: AFFF, also known as "light water," is a synthetic, film-forming foam designed for use in shipboard firefighting systems.

AFFF DISTRIBUTION SYSTEM: The AFFF distribution systems installed on the SSN-21 class submarines is designed to supply AFFF to sprinklers that protect the diesel generator space, the lube oil bay bilge, the areas outboard of the fan room, and the area around hydraulic plants in the engine room. The AFFF system also provides AFFF to all 1- ½ inch seawater fireplugs. AFFF is generated from two AFFF distribution systems installed forward and aft.

AFFF FIRE EXTINGUISHER (AFFF): Used to provide a vapor seal over a small fuel spill, to extinguish small Class B fires (such as deep-fat fryers), and for standing fire watch during hot work.

AFFF SINGLE-SPEED INJECTION PUMP: Permanently mounted, positive displacement, electrically driven, sliding-vane type of pump.

AFFF SPRINKLER SYSTEM: An AFFF sprinkler system is a subsystem of AFFF generating systems.

AFFF STATION OPERATOR: The AFFF station operator ensures that there is a constant supply of AFFF to the hose team for firefighting.

AFFF TANKS: The tanks are rectangular or cylindrical in shape and are fabricated out of 90/10 copper-nickel or corrosion-resistant steel. Each service tank is located inside the AFFF station and is fitted with a gooseneck vent, drain connection, fill connection, liquid level indicator, recirculating line, and an access manhole for tank maintenance.

AFFF TRANSFER PUMPS: Permanently mounted, single-speed, centrifugal type, electrically driven pump. These pumps are provided in 360 GPM capacities. The transfer pump moves AFFF concentrate through the AFFF fill-and-transfer subsystem to all AFFF station service tanks on a selective basis.

AFFF TRANSFER SYSTEM: The transfer system can deliver AFFF concentrate to on-station service tanks via a transfer main. The transfer main consists of a large pipe with smaller branch connections interconnecting the AFFF service and storage tanks.

AFFF TWO-SPEED INJECTION PUMP: Designed to meet the demand for either a low or a high firefighting capability. The two-speed AFFF pump consists of a positive displacement pump rated at 175 psi, a motor, and a reducer, coupled together with flexible couplings and mounted on a steel base.

AFLOAT TRAINING GROUPS (ATG): Provide Navy ships with examples and packages of recommended damage control drills.

ALPHA (A) FIRES: Those that occur in such ordinary combustible materials as wood, cloth, paper, upholstery, and similar materials.

ANTIFLASH CLOTHING: Intended to protect personnel from transient high temperatures that may occur from the use of high explosive weapons and from being burned in a fire. Antiflash clothing consists of an antiflash hood and antiflash gloves.

AQUEOUS POTASSIUM CARBONATE (APC): Are installed in Navy ships to provide protection for galley deep-fat and doughnut fryers and their exhaust systems. Aqueous potassium carbonate is specifically formulated to extinguish fire in the reservoirs by combining with the hot cooking oil surface to form a combustion-resistant soap layer, thereby cutting off the grease from its source of oxygen.

ATMOSPHERIC TESTING: The process of using gas analyzers to check the atmosphere for oxygen, toxic, and combustible gases. The testing is conducted by the GFE, GFEA, or GFEPO. The level of oxygen must be between 19.5 and 22 percent. Combustible gases must be less than 10 percent of the lower explosive limit and all toxic gases must be below their threshold limits before the compartment is certified safe for personnel without breathing apparatuses. See GAS FREEING.

AUTOGENOUS IGNITION TEMPERATURE: The temperature just adequate to cause the vapors from a petroleum product to burst into flames without the application of a spark or flame.

AUTO-IGNITION POINT: The minimum temperature required to initiate self-sustained combustion of a substance independent of external ignition sources of heat.

AUTO-IGNITION/SELF-IGNITION POINT: The lowest temperature to which a substance must be heated to give off vapors that will burn without the application of a spark or flame.

BACKDRAFT: An explosion that results from combining fresh air with hot flammable fire gases when they have reached their auto-ignition temperatures.

BALANCING VALVE: Automatically proportions the correct amount of AFFF concentrate with seawater. The balancing valve is a diaphragm-actuated control valve that responds to pressure changes between the AFFF concentrate supply line and the firemain.

BATTLE BILL: The ship's Battle Bill is tailored to your ship for battle organization. You may need to provide information to the operations department when it is updated.

BATTLE DRESSING STATIONS (BDS): Most ships have a minimum of two battle dressing stations equipped for emergency handling of personnel battle casualties. Each battle dressing station must be accessible to the stretcher-bearers from repair parties within the vicinity. Medical department personnel as assigned by the senior member of that department should man each battle dressing station.

BRAVO (B) FIRES: Those that occur in the vapor-air mixture over the surface of flammable liquids, such as gasoline, jet fuels, diesel oil, fuel oil, paints, thinners, solvents, lubricating oils, and greases.

BUFFER ZONE: The area between the inner and outer smoke boundaries established for a Class B fire in a machinery space.

BULKHEADS: Vertical walls that run both transversely and longitudinally through the interior of a ship which divide it into compartments.

CARBON DIOXIDE (CO_2): An effective agent for extinguishing fires by smothering them and also produced by a fire when there is complete combustion of all of the carbon in the burning material. v is a colorless and odorless gas.

CARBON DIOXIDE (CO_2) FIRE EXTINGUISHER: Used on small electrical fires (Class C) and has limited effectiveness on Class B fires.

CARBON DIOXIDE (CO_2) HOSE-AND-REEL SYSTEM: Consists of two cylinders, a length of special CO_2 hose coiled on a reel, and a horn-shaped nonconducting nozzle equipped with a second control valve.

CARBON MONOXIDE (CO): A colorless, odorless, tasteless, and nonirritating gas. However, it can cause death even in small concentrations.

CHARLIE (C) FIRES: Fires that occur in electrical wiring or equipment.

COLD WEATHER BILL: The Cold Weather Bill is used to prepare the ship for cold weather operations.

COMBUSTION: A rapid chemical reaction that releases energy in the form of light and noticeable heat. Most combustion involves rapid OXIDATION, which is the chemical reaction by which oxygen combines chemically with the elements of the burning substance.

COMMUNICATOR: The communicator may function as a phone talker or messenger depending on the ship's firefighting and emergency response organizational structure and communications equipment configuration. They must be able to operate the designated DC communications circuits. They may use a computer system, wireless free communications radios (WIFCOM), ships telephone system, sound-powered phones, or written messages.

COMPARTMENT VENTING: Using a weather deck area directly above any portion of the affected compartment to cut a one-foot square hole in the deck to allow heat and gases to escape from the affected compartment.

CONCENTRATION: The quantity of a substance per unit volume. Examples of concentration units are milligrams per cubic meter (mg/m3); parts per million (ppm) for vapors, gases, fumes, or dusts; fibers per cubic centimeter (fibers/cc) for vapors or gases.

CONDUCTION: Transfer of heat through a body or from one body to another by direct physical contact.

CONFLAGRATION (CONFLAG): A large magnitude of damage caused by collisions, explosions, or aircraft crashes fires that leads to fires, flooding, structural damage, equipment malfunctions, personnel injuries, and death.

CONFLAG STATION: A protected space within a ship with a view of aircraft hangar areas to which a watch stander is assigned for the purpose of monitoring the assigned hangar area for fire hazards and fires as well as the response to such with regard to the operation of lighting, door, AFFF overhead sprinkling, and communications system controls contained within the subject station.

CONVECTION: Transfer of heat through the motion of circulating gases or liquids.

DAMAGE CONTROL ASSISTANT (DCA): The DCA is the primary assistant to the damage control officer in the areas of damage control; firefighting; and chemical, biological, and radiological defense.

DAMAGE CONTROL BOOKS: These books contain descriptive information, tables, and diagrams. Each book is pertinent to an individual ship. The information given covers the following six subjects: "Damage Control Systems," "Ship's Compartmentation," "Ship's Piping Systems," "Ship's Electrical Systems," "Ship's Ventilation Systems," and "General Information."

DAMAGE CONTROL CENTRAL (DCC): The primary purpose of DCC is to collect and compare reports. Location onboard ship where damage control operations are coordinated through and also where direction is given to repair teams.

DAMAGE CONTROL SUPERVISOR (DCS): The DCS is personnel that stand hourly shifts at DCC during normal ship operations and emergencies. The DCS track the status of fire protection systems on a control panel or computer system. The DCS will assist the DCA during shipboard emergencies.

DAMAGE CONTROL REPAIR STATION (DCRS): The DCRS is an area designated as a locker. It contains equipment for firefighting and battle damage repair.

DAMAGE CONTROL ORGANIZATION: The damage control organization consists of two elements—the damage control administrative organization and the damage control battle organization.

DAMAGE CONTROL TRAINING: Consistent training produces an optimal level of readiness that prepares members of repair party teams to react more efficiently and effectively to actual casualties.

DAMAGE CONTROL TRAINING TEAM (DCTT): Composed of qualified senior members of the ship's crew specifically tasked to ensure the ship's company maintains the highest level of battle readiness. This training is maintained through comprehensive training programs, which include lectures and drill scenarios.

DCTT TEAM LEADER: The executive officer serves as the chairman of the planning board for training and team leader of the DCTT. The executive officer will coordinate the planning and execution of the ship's training effort. The team leader of the DCTT is responsible for the management of the training team.

DCTT TEAM COORDINATOR: The ship's senior Damage Controlman or Hull Maintenance Technician normally holds the position of DCTT team coordinator.

DECAY STAGE: The stage whenever all the available fuels and combustibles in the compartment are consumed. Once all the materials are consumed, the combustion rate slows down until the fire goes out.

DELTA (D) FIRES: Those that occur in combustible metals, such as magnesium, titanium, and sodium.

DESMOKING: The process of removing smoke and combustible gases from a compartment after fire is extinguished. Smoke and combustible gases can be removed using potable blowers such as the Ramfan, box fan, or installed ventilation fans.

DEWATERING: The process of removing flooding water or hazardous liquid from the ship by using equipment such as portable eductors, P-100 pump, electrically submersible pumps and installed eductors.

DIRECT FIRE ATTACK: A method of attacking a fire in which fire fighters advance into the immediate fire area. The extinguishing agent is applied directly onto the seat of the fire to extinguish the fire or spray a water fog (fog attack) into the hot gas layer over the seat of the fire to gain control.

DRY CHEMICAL EXTINGUISHER: Used primarily on Class B fires. PKP is the chemical most often used in these extinguishers.

ELECTRIC SUBMERSIBLE PUMP (ESP): The electric submersible pump (ESP) is a portable centrifugal pump driven by a water-jacketed constant speed ac electric motor. It is designed to remove flood water from ships and submarines. It operates as single-phase or three-phase to deliver 140 gpm against a maximum head of 70 feet and 180 gpm at a 50-foot static head.

EMERGENCY AIR BREATHING (EAB) SYSTEMS: The emergency air breathing (EAB) system is designed to supply oxygen to personnel conducting emergency operations in a toxic atmosphere such as smoke, vapors, fumes, or gases.

EXPLOSION SUPPRESSION FOAM (ESF): Explosion suppression foam (ESF) is a flexible polyurethane foam material installed in certain aircraft fuel tanks and cells that provides protection against explosion.

EXPLOSIVE RANGE: A scale that indicates the explosive nature of gases or vapors. The relationship of the concentration of the vapor present; its temperature and pressure is expressed as a percent by volume in air. If the explosive range falls below the lower explosive limit (LEL), the mixture of air and vapor is too lean for an explosion. If the explosive range is above the maximum explosive range or upper explosive limit (UEL), the mixture of vapor and air is too rich to be explosive.

EXPLOSIVE-PROOF: Describes an apparatus, device, or equipment that is tested and approved for use in hazardous atmospheres, as defined in the National Electrical Codeã. Explosive-proof devices are designed to withstand internal explosions and prevent hot vapors or particles from exiting before they become significantly cooled.

FIRE BOUNDARYMEN: The fire boundarymen set primary and secondary fire boundaries as directed by the repair party leader or fire marshal. They secure all doors, hatches, and openings in the boundary of the fire area. They remove or relocate combustibles as required. They cool boundaries with hoses as required. They are normally monitored by and report to the roving investigators.

FIREFIGHTER'S ENSEMBLE: Designed to protect the fire fighter from short duration flame exposure, heat, and falling debris. The components of the firefighter's ensemble include the firefighter's coveralls, antiflash hood, damage control/firefighter's helmet, firefighter's gloves, and firefighter's boots.

FIRE HOSE STATION: A fire hose station is the location where fireplug and associated equipment are stored; commonly referred to as either a fire station or a fireplug.

FIRE POINT: The temperature at which a fuel will continue to burn after it has been ignited.

FIRE PROTECTION ENGINEERING: Fire protection engineering is applying science to engineering principles to protect people, property, and their environments from the harmful and destructive effects of fire and smoke through fire detection, suppression and mitigation.

FIRE SAFETY ENGINEERING: Fire safety engineering is another element of fire protection engineering which focuses on human behavior and maintaining a dependable environment for evacuation from a fire.

FIRE TRIANGLE: Three components are heat, fuel, and oxygen. Fires are generally controlled and extinguished by eliminating one side of the fire triangle. If any of the fuel, heat, or oxygen is remove, it will prevent or extinguish a fire.

FIRE TETRAHEDRON: An uninhabited chain reaction that is a fourth component added to the fire triangle. Fires are generally controlled and extinguished by eliminating one side of the fire triangle but in this case the fourth side. If any of the fuel, heat, oxygen or uninhabited chain reaction, the fire will be extinguish.

FIRE UNDER CONTROL: A fire under control is when one or more hose teams are making progress advancing on a fire and the fire is contained in a single area within a compartment.

FIREMAIN SYSTEM: Receives water pumped from the sea. It distributes this water to fireplugs, sprinkling systems, flushing systems, machinery cooling-water systems, washdown systems, and other systems as required. The firemain system is used primarily to supply the fireplug and the sprinkling systems; the other uses of the system are secondary.

FIRE MARSHAL: The fire marshal is an assistant to the CHENG and aids the DCA with training the crewmembers on fire prevention, firefighting, emergency response and battle damage repairs. The fire marshal is normally the most senior fire protection engineer on the ship except for aircraft carriers. In this case the fire marshal is a limited duty officer or warrant officer with an extensive background in firefighting and emergency response.

FLASHOVER: A flashover is the transition from a growing fire to a fully developed fire in which all combustible items in the compartment are involved in fire.

FLASH POINT: The lowest temperature at which a flammable substance gives off vapors that will burn when a flame or spark is applied.

FLIGHT DECK: In aircraft carriers the uppermost complete deck. It is the deck from which aircraft take off and land.

FOG FIRE ATTACK: Fog fire attack is a method of attacking a fire in which fire fighters outside the fire area discharge water fog into the overhead area of the fire to reduce the affected compartment temperatures, radiant heat, and flaming combustion.

FOREIGN OBJECT OF DESTRUCTION (FOD): Objects and debris that can cause damage to aircraft engines.

FRESHWATER HOSE REEL SYSTEM: The freshwater hose reels are installed on ships to provide a rapid response firefighting capability for Class A fires.

FUEL: A solid, liquid, or even a vapor. Some of the fuels you will come into contact with are rags, paper, wood, oil, paint, solvents, and magnesium metals.

FULLY-DEVELOPED FIRE STAGE: The stage where all flammable materials in the compartment have reached their ignition temperature and burning. The rate of combustion will normally be limited by the amount of oxygen available in the air to provide combustion.

FUME: Solid particles formed by condensation of metals from the gaseous state.

GALLERY DECK: First deck or platform below the flight deck.

GAS FREEING: Operations performed in testing, evaluating, removing, or controlling hazardous materials or conditions within or related to a confined space which may present hazards to personnel entering or working in or adjacent to the space.

HALON: Halon is a halogenated hydrocarbon that provides non-flammability and fire extinguishing properties that is an effective agent against Class A, Class B, and Class C fires. Halon extinguishes fires by interrupting the chemical chain reaction of the fire.

HANGAR DECK: The deck on which aircraft are stowed and serviced when not on the flight deck.

HEAT: Involves three methods-conduction, convection, and radiation.

HEAT CASUALTY: An individual unable to perform his or her duties as a result of heat exhaustion or heat stroke.

HEAT EXHAUSTION: A physical condition caused by exposure to high temperature combined with physical exertion, and marked by faintness, nausea, and profuse sweating; can be considerably reduced by proper physical conditioning and increased fluid intake.

HEAT STRESS: Heat stress is a pathological condition in which the body's cooling mechanisms are unable to dissipate the heat load generated.

HEAT STROKE: A state of collapse or prostration, usually accompanied by high fever, brought on by exposure to heat; has a 50 percent mortality rate but accounts for only a small percentage of heat casualties.

HEPTAFLUOROPROPANE (HFP): HFP is the Navy's term for a specific gaseous fire extinguishing agent which is an alternative to halon 1301 in some of the newer ships. HFP consists of several compounds such as carbon, fluorine, and hydrogen in the formula C_3F_7H.

HOSE CONTROL DEVICE: A device used on large air-capable ship platforms to assist with cooling ordnance. It is designed to be securely fastened to the deck and connected to a navy standard firefighting nozzle and hose to direct the flow of AFFF or water towards any exposed or at-risk ordnance during a fire on the flight deck or hangar deck.

HOSEMAN: A hoseman runs the attack hose from the fireplug to the scene, and you will keep the hose from getting fouled while fighting the fire and relay spoken messages and orders between the on-scene leader and the nozzleman.

HYCHECK VALVE: Diaphragm type, fail open, seawater pressure-operated control valve, which allows the flow of seawater from the firemain system to be mixed with AFFF concentrate.

HYDRAZINE: Hydrazine is a clear, oily, water-like liquid that smells like ammonia. Hydrazine is used as fuel for the F-16 aircraft's emergency power unit (EPU).

HYDROCARBON: A compound containing only carbon and hydrogen. Hydrocarbons are the principal constituents.

HYDROGEN SULFIDE (H_2S): Generated in some fires. It is also produced by the rotting of foods, cloth, leather, sewage, and other organic materials.

HYTROL VALVE: Diaphragm type, fail open, seawater pressure-operated control valve that controls the flow of AFFF solution to systems.

IGNITION: The act or action of causing a substance to burn; the means whereby a material starts burning.

INDIRECT FIRE ATTACK: Indirect fire attack is a method of attacking a fire in which fire fighters outside the fire area discharge water fog into the fire area through a cracked open door or a bulkhead or overhead penetration.

IN-PORT EMERGENCY TEAM (IET): The in-port fire party will function as a repair party while the ship is in port. CBR defense operations are not a normal evolution for an in-port fire party.

INVESTIGATOR—Investigators are assigned to repair lockers to ensure that no further damage occurs outside the boundaries of the existing casualty. Investigators normally operate in pairs, traveling assigned routes and reporting conditions to the repair locker.

JETTISON: Moving an object from the ship by tossing it in the sea.

JP-5: A high flash point, kerosene-type aircraft turbine fuel, specifically designed for storage and use on naval ships.

LOWER EXPLOSIVE LIMIT (LEL): The minimum percent by volume of a gas that, when mixed with air at normal temperature and pressure, will form a flammable mixture.

LOWER FLAMMABLE LIMIT (LFL): The minimum concentration of a combustible gas or vapor in air, usually expressed in percent by volume at sea level, which will ignite if a sufficient ignition source of energy is present.

MACHINERY SPACE: Machinery space is main and auxiliary machinery spaces that contain any of the following: installed firefighting systems, oil-fired boilers, internal combustion engines, gas turbines, or steam turbines.

MAGAZINE SPRINKLER SYSTEMS: Sprinkler systems are used for emergency cooling of, and fire fighting in, magazines, ready-service rooms, ammunition, and missile handling areas. A magazine sprinkler system consists of a network of pipes. Magazine sprinkler systems can completely flood their designated spaces within an hour.

MESSENGER: Individual responsible to relay orders and information. These messages will normally be relayed between the scene, the repair locker, and, if in port, the quarterdeck.

MISCELLANEOUS SEAWATER SPRINKLER SYSTEMS: Miscellaneous seawater sprinkler systems are normally installed in spaces where the quantity and combustibility of materials present is high enough that, should a fire occur in these materials, hose line attack would not succeed in preventing compartment burnout and major damage.

MISSILE GAS SYTEM: The missile gas system is fire suppressing system that is installed in submarine's missile launching tube. The system uses nitrogen to pressurize the launching tube during missile launches or fail launches.

MOBILE FIREFIGHTING VEHICLE (MFFV): A mobile vehicle used to firefighting on large air-capable ships. The A/S32P-25 or P-25 for short is a diesel-driven vehicle that is equipped with a turret, AFFF hand line, 55-gallon AFFF tank, 750-gallon water tank, proportioning pump, three portable halon 1211 extinguishers, and nursing connections.

MOGAS: Combat automotive gasoline that has a low octane rating that may cause knocking in engines. The relative amount of lead influences the octane rating.

NAVAL SHIPS' TECHNICAL MANUAL (NSTM): These manuals cover different aspects of damage control, which include the following: firefighting, flooding, ship's stability, and CBR countermeasures. Study of the NSTMs will help you complete your damage control personnel qualification standards.

NOZZLEMAN: The nozzleman mans the attack hose nozzle so that the fire may be extinguished.

NURSING: Resupplying the MFFVs with water or AFFF to continue to fight the fire at the scene.

ON-SCENE LEADER (OSL): The on-scene leader is the person in charge at the scene of the fire or casualty.

OVERHAUL: An examination and cleanup operation. It includes finding and extinguishing hidden fire and hot embers and determining whether the fire has extended to other parts of the ship.

OXIDIZING MATERIAL: A chemical compound that spontaneously releases oxygen at normal temperature and air pressure or under slight heating.

OXYGEN: The content of the surrounding air. Ordinarily, a minimum concentration of 15 percent oxygen in the air is needed to support flaming combustion.

PROXIMITY FIRE FIGHTING PROTECTIVE ENSEMBLE (PFFPE): Multiple elements of compliant protective clothing and equipment that when worn together provide protection from fires on aircrafts.

P-100 PORTABLE PUMP: The P-100 portable pump is an engine-driven centrifugal pump assembly that uses diesel or JP-5 fuel. An air-cooled single-cylinder, four-cycle diesel engine rated at 10 horsepower. The P-100 is used for firefighting and removing flood water.

PHONE TALKER: The phone talker mans the phone between the supervisor at their location and other stations and receives messages from other phone talkers and relays them to their supervisor.

PLUGMAN: The plugman connects the hose to the fireplug, and when directed to do so and while the nozzle is closed, open the fireplug valve to activate the hose.

PORTABLE AFFF INJECTION UNIT (PAIU): The PAIU is a three gallon cylinder designed to store and deliver 1% AFFF concentrate to the AFFF/SW system. The PAIU injects AFFF concentrate at quick disconnect tee fittings at each fire station and fire hose reel.

POST-FIRE INVESTIGATION: An investigation to determine the point of origin, types of combustibles involved, path of fire spread, ignition source, and significant events in the growth and eventual extinguishment of the fire. The investigation will be directed toward recreating the conditions that caused the fire and identifying any changes in design or procedures that could have prevented the fire or lessened its spread and intensity.

POWERCHECK VALVE: Diaphragm type, normally closed, seawater pressure-operated control valve. This valve allows the flow of AFFF from the pump to be mixed with seawater and protects the AFFF tank from seawater contamination or dilution.

POWERTROL VALVE: Diaphragm type, normally closed, seawater pressure-operated control valve. This valve allows the flow of AFFF/seawater solution through the distribution system or controls seawater flow on flight deck injection systems.

PURPLE-K-POWDER (PKP): Potassium bicarbonate ($KHCO_3$) powder used to extinguish Class B and Class C fires.

QUANTAB CHLORIDE TITRATOR V: Quantab chloride titrator strips are used to measure salt (chloride) in aqueous solutions.

REFLASH WATCH: Once a fire is extinguished, the reflash watch is set by a member of the fire hose team who remains near the seat of the fire with a charged hose and observes the area to ensure that no new fire breaks out.

REFRACTOMETER: Gives accurate readings of total dissolved solids in aqueous solutions.

REPAIR PARTIES: Qualified shipboard personnel responsible for executing damage control duties in a training or actual damage control situation.

REPAIR PARTY MANUAL: The repair party manual provides detailed information on the standard methods and techniques used in damage control as outlined in *NWP* 3-20.31.

RESCUE AND ASSISTANCE BILL: The Rescue and Assistance Bill organizes qualified personnel by duty section or the entire ship to render emergency assistance outside the ship.

ROLLOVER: A sudden spread of flame through the unburnt gases and vapors in the upper layer across the overhead of a space.

SELF-CONTAINED BREATHING APPARATUS (SCBA): Type of respirator that allows the user complete independence from a fixed source of air.

SHIP INFORMATION BOOK: When a ship is built for the Navy, the builder prepares a ship information book (SIB). The ship's crew uses the SIB to familiarize themselves with the ship's characteristics.

SINGLE MAIN FIREMAIN SYSTEM: Consists of a single piping run that extends fore and aft. This type of firemain is generally installed near the centerline of the ship, extending forward and aft as far as necessary.

SMOKE CONTROL ZONE: The area between the inner and outer smoke boundaries established for fires that involve primarily Class A or Class C materials.

SOLENOID-OPERATED PILOT VALVE (SOPV): Electrically operated pilot valves that control the activation of many AFFF fire-extinguishing systems. All SOPVs (master and service) have four control line ports; one port is always connected to supply pressure (firemain), and a second port is the valve drain (which should be piped to discharge within the coaming of the AFFF station).

SSN-774 CLASS FIREMAIN SYSTEM: The SSN-774 class firemain is a firemain system installed on SSN-774 class submarines consisting of two firemains located forward and aft sections of the submarine. AFFF solution can be injected into this firemain system and distribute AFFF to fire hose stations, hose reels, and the diesel overhead sprinkler system.

SSN-21 CLASS FIREMAIN SYSTEM: The SSN-21 class firemain system is a firemain system installed on SSN-21 class submarines consisting of one firemain system installed in forward compartment and another in the after compartment of the submarine. The firemain is supplied with sea water from the pressurized trim tanks and distributes sea water to services as required.

STRETCHER-BEARER: The stretcher-bearer is required to take the repair locker first-aid kit, or box, to or near the scene. If medical department personnel are available, they will help them in administering first aid, as required.

TOXIC GAS BILL: The Toxic Gas Bill specifies the procedures and assigns duties and responsibilities for controlling and minimizing toxic gas casualties.

TRIM SYSTEM: A firemain system installed on submarines consisting of pumps, piping, tanks, and valves through which seawater is supplied to fire hose stations, AFFF systems, and sprinkling systems that provide water for flooding the pyrotechnics, small arms ammunition, and chlorate candle stowage lockers.

UPPER EXPLOSIVE LIMIT (UEL): Upper end of the explosive range. Concentrations above this limit are too rich to explode or burn. Concentrations below the LTEL are within the explosive range.

VENTILATING SYSTEMS: The ventilating system consists of vent ducting and fan units that circulate a clean air conditioned atmosphere throughout the ships and submarine. The ventilating system can be used to remove smoke and toxic gases from ships and submarines.

WATER MIST SYSTEMS: Water mist is a fire protection agent which replaces halon 1301 in new ship designs. Water mist for machinery spaces is a total-space fire extinguishing system which discharges high-pressure (approximately 1000 psi) fresh water as a fine mist from nozzles located in all levels except the bilge. High-pressure water mist is effective at suppressing oil pool fires, oil spray fires and class.

WEATHER DECK: A deck or part of a deck exposed to the weather.

Appendix II – Abbreviations and Acronyms

ABT – automatic bus transfer

ACHO – Aircraft handling officer

ADTT – air department training team

AFFF – aqueous film-forming foam

AMMO – arms ammunition

APC – aqueous potassium carbonate

ATG – Afloat Training Group

ATT – aviation training team

BDS – battle dressing station

CHENG – chief engineer

CBRN – chemical, biological, radiological, and nuclear

CBRN-D – chemical, biological, radiological, and nuclear defense

CDO – command duty officer

CFM – cubic feet per minute

CO – carbon monoxide

CO – commanding officer

CO_2 – carbon dioxide

CONFLAG – conflagration

CVA – attack aircraft carrier

CVN – nuclear-powered aircraft carrier

DCA – damage control assistant

DCC – damage control central

DCO – damage control officer

DCPO – damage control petty officer

DCREL – damage control rescue and assistance reentry locker

DCRS – damage control repair station

DCS – damage control supervisor

DCTT – damage control training team

DCUL – damage control unit lockers

DCUPS – damage control unit patrol stations

DDG – guided-missile destroyer ship

EAB – emergency air-breathing system

EEBD – emergency escape breathing device

ESF – explosive suppressant foam

ESP – electric submersible pump

FFE – firefighter's ensemble

FOD – foreign object of destruction

GFE – gas free engineer

GFEA – gas free engineer assistant

GFEPO – gas free engineer petty officer

GPM – gallons per minute

GQ – general quarters

H_2S – hydrogen sulfide

HBr – hydrogen bromide

HCl – hydrogen chloride

HCN – hydrogen cyanide

HCO – helicopter control officer

HFP – heptafluoropropane

HF – hydrogen fluoride

IET – in-port emergency team

IWO – integrity watch officer

LCAC – air cushion landing craft

LCS – littoral combat ship

LEL – lower explosive limit

LHA – landing helicopter assault ship

LHD – landing helicopter deck ship

LOT – lockout trunk

LOX – liquid oxygen

LPD – landing platform dock ship

LSD – dock landing ship

MCM – mine countermeasures ship

MFFV – mobile firefighting vehicle

MOS – military occupational specialty

NAVEDTRA – naval education training

NAVSEA – Naval Sea Systems Command

NEC – navy enlisted code

NFTI – naval firefighter's thermal imager

NRL – naval research laboratory

NSTFP – navy standard fire pump

NSTM – Naval Ships' Technical Manual

OHO – ordnance handling officer

OJT – on-the-job training

OSL – on-scene leader

OOD – officer of the deck

PAIU – portable AFFF injection unit

PEARS – portable electrical access and rescue system

PECU – portable exothermic cutting unit

PFFPE – proximity firefighting protective ensemble

PKP – purple-K-powder

PYRO – pyrotechnics

PPM – parts per million

RPM – rotations per minute

RPL – repair party leader

SCBA – self-contained breathing apparatus

SFPE – *society of fire protection engineers*

SOPV – solenoid-operated pilot valve

SSN – nuclear-powered general purpose attack submarine

SSBN – nuclear-powered ballistic missile submarine

TIC – thermal imager camera

UEL – upper explosive limit

WIFCOM – wire-free communication

XO – executive officer

Appendix III – Index

A

accessing the Space 185
ACHO 206, 208, 209, 210, 219, 221, 260
administrative Organization 28
ADTT 210, 222, 223, 243, 265, 283
AFFF 8, 11, 54, 55, 58, 67, 74, 75, 77, 80, 82, 83, 84, 85, 86, 95, 96, 103, 106, 110, 111, 112, 113, 114, 115, 116, 117, 118, 119, 120, 121, 122, 123, 124, 137, 147, 152, 153, 154, 155, 156, 157, 158, 159, 165, 170, 172, 178, 179, 182, 184, 185, 193, 198, 200, 201, 202, 207, 209, 212, 214, 215, 217, 219, 221, 222, 227, 228, 229, 230, 231, 232, 233, 234, 235, 236, 237, 238, 241, 242, 244, 245, 249, 250, 251, 253, 254, 255, 256, 257, 259, 260, 265, 266, 267, 269, 275, 276, 277, 278, 279, 280, 281, 283, 286
AFFF distribution system 155, 265
AFFF eductor system 110
AFFF flight deck and deck edge sprinkling system 235
AFFF flight deck weapons staging sprinkling system 235
AFFF hangar deck sprinkling systems 235
AFFF injection systems 234
AFFF proportioning 215, 234
AFFF single-speed injection pump 110, 111
AFFF station operator 58
AFFF two-speed injection pump 110, 112
afloat Training Group 15, 283
air boatswain 208, 210, 219, 229, 254
air-capable ships 44, 79, 205, 206, 207, 212, 225, 233, 235, 237, 238, 241, 243, 244, 277
aircraft brake and wheel assembly fires 247
aircraft crash, salvage, and rescue officer (air boatswain) 206
aircraft crash, salvage, and rescue supervisor 206
aircraft debris and running fuel fires 246, 255
aircraft electrical and electronic fires 251
aircraft engine wet 251
aircraft handling officer (ACHO) 206
aircraft tailpipe fires 249
air department training team 206
air gunner 206
air officer (air boss) 206

anti-flash clothing 93
APC 75, 80, 140, 141, 142, 143, 144, 147, 153, 161, 162, 165, 266, 283
aqueous film-forming foam 8
aqueous potassium carbonate 75, 80
atmospheric testing 201
at-sea fire party 46
attacking the fire 188
attack team leader 52
automatic wet type miscellaneous sprinkler systems 147
aviation fuel officer 206
aviation fuel repair team 44, 212

B

back draft 176
background assistance detail 212, 216, 217, 218
background assistance leader 216, 217, 218
balanced-pressure proportioner (type ii) 110, 113
balanced-pressure proportioner (type iii) 110, 114
batteries 94, 216, 226, 230, 231, 246
battle dressing stations 20, 45
battle organization 23, 28, 33, 34, 35, 60, 168, 267, 270
BDS 20, 33, 34, 35, 36, 40, 45, 59, 267, 283
blowers 57, 95, 98, 200, 201, 271
boundaryman 56
box fan 98, 100, 200, 271
breathing air management 183

C

carbon dioxide 75, 77, 86
carbon dioxide (CO_2) hose and reel systems 126
carbon monoxide 70, 98, 102, 200, 201, 231, 283
casualty coordinator 150
cdo 22, 23, 283
cheng 17, 19, 20, 21, 273, 283
chief engineer 17, 283
class a fire 68, 137, 169, 171
class a fires 67, 76, 77, 78, 83, 85, 86, 137, 161, 169, 274
class alpha fire 67
class b fires 67, 80, 82, 84, 86, 96, 110, 123, 170, 173, 179, 198, 201, 228, 265, 268, 271
class bravo fire 67
class c fire 171, 200
class charlie fire 68
class delta fire 68
class d fires 171, 172
CO_2 68, 70, 74, 75, 77, 78, 82, 83, 84, 86, 87, 124, 125, 126, 127, 130, 131, 133, 134, 135, 136, 147, 158, 171, 201, 216, 221, 230, 231, 237, 238, 249, 250, 251, 252, 268, 283
combustion 3, 63, 64, 65, 66, 68, 70, 72, 75, 78, 82, 124, 128, 141, 162, 172, 175, 191, 192, 227, 228, 267, 268, 271, 274, 277, 278
command duty officer 22
commanding officer 16
communicator 59, 218, 268
compartment venting 197
composite materials 226, 232
composite system 103, 104

conduction 73, 176, 275
conflag station 219, 236, 260
conflag station operator 219, 260
convection 72, 73
crash 37, 39, 41, 45, 205, 206, 208, 210, 211, 212, 215, 217, 218, 220, 221, 222, 227, 229, 230, 231, 232, 233, 235, 238, 239, 240, 241, 242, 243, 244, 246, 247, 248, 249, 252, 253, 254, 255, 256, 257, 260, 261, 265
crash and salvage team 44
crash, salvage, and rescue team 211, 220, 221, 229, 230, 238, 239, 240, 241, 242, 243, 244, 247, 248, 249, 252, 254, 257, 261

D

damage control assistant 9, 284
damage control book 11, 20
damage control central 12, 35, 134
damage control officer 17, 269, 284
damage control petty officer 30
damage control repair station 18, 37, 39, 40, 41, 42, 43
damage control repair station leader 50
damage control rescue and assistance reentry locker 44
damage control supervisor 31
damage control training team 15, 23
damage control unit locker 43
damage control unit patrol station 44
DCA 9, 13, 15, 19, 21, 22, 29, 30, 31, 32, 35, 49, 50, 58, 59, 89, 150, 177, 179, 181, 183, 197, 202, 207, 269, 270, 273, 284
DCC 12, 13, 20, 31, 34, 35, 36, 42, 49, 50, 59, 103, 105, 140, 147, 150, 158, 178, 270, 284
DCO 17, 284
DCPO 20, 30, 82, 84, 86, 88, 284
DCREL 44, 284
DCRS 18, 19, 20, 33, 34, 35, 36, 37, 40, 41, 42, 43, 44, 45, 46, 48, 49, 50, 51, 52, 56, 57, 58, 59, 60, 84, 87, 178, 179, 181, 182, 183, 184, 260, 270, 284
DCS 31, 270, 284
DCTT coordinator 23, 24
DCTT leader 23, 24, 25
DCTT team members 23, 25
DCUL 43, 44, 284
DCUPS 44, 284
decay stage 174, 175
department heads 29, 30
desmoking 57, 98, 99, 100, 178, 194, 200, 201, 202
desmoking 100, 200
dewatering 202
direct attack 188, 189, 192
division officer 29
dry chemical extinguisher 82
dynamics of fire 172

E

EAB 153, 163, 271, 284
EEBD 89, 90, 284
electrician 56, 180
electric motor-driven fan 98
emergency air breathing 163, 271
emergency escape breathing device 89
engineer officer 17, 150
executive officer xo 17
explosion suppression foam 231, 272
extinguishment 198

F

fans 57, 98, 99, 136, 200, 271
FFE 90, 182, 284
fire alarm system 102
fire classifications 67
fire components 63
firefighter's ensemble 69, 90, 197, 239, 241, 272, 284
firefighter's gloves 93
firefighting ix, x, xi, xii, xiii, 4, 5, 6, 7, 8, 9, 10, 11, 12, 13, 15, 16, 17, 18, 19, 20, 21, 22, 23, 24, 26, 27, 28, 29, 30, 32, 33, 34, 35, 36, 40, 41, 42, 43, 44, 46, 47, 50, 51, 52, 53, 54, 55, 56, 58, 59, 60, 61, 62, 67, 69, 71, 72, 74, 75, 76, 77, 79, 80, 81, 88, 89, 91, 93, 95, 100, 101, 102, 106, 107, 108, 109, 112, 113, 120, 123, 124, 129, 147, 149, 150, 151, 152, 153, 161, 163, 164, 167, 168, 169, 172, 174, 175, 176, 177, 178, 179, 180, 181, 182, 183, 184, 185, 186, 188, 189, 190, 191, 192, 193, 194, 196, 197, 198, 199, 202, 203, 205, 206, 207, 208, 209, 211, 212, 213, 214, 215, 217, 218, 220, 221, 222, 223, 225, 226, 227, 228, 231, 232, 233, 234, 235, 236, 238, 239, 240, 241, 242, 243, 244, 246, 248, 251, 252, 253, 254, 255, 256, 257, 260, 261, 265, 266, 268, 269, 270, 273, 274, 275, 277, 278, 285, 286
firefighting strategies 168
firefighting team considerations 177
firefighting teams 72, 150, 211
fire growth 174
fire hose station 106, 107, 158, 160, 272
firemain system 12, 76, 98, 99, 103, 104, 105, 106, 107, 108, 109, 118, 153, 154, 167, 238, 273, 276, 281
fire marshal xi, 21, 22, 30, 44, 46, 49, 56, 177, 179, 272, 273
fireplugs 76, 103, 107, 154, 155, 265, 273
fire properties 169
fire protection engineer 2, 167
fire pump 105, 285
fire spread 176
fire tetrahedron 63
fire triangle 63, 66, 72, 74, 80, 169, 273
flame 69
flashover stage 174
flight deck ix, 19, 45, 58, 74, 83, 87, 117, 121, 207, 208, 209, 210, 211, 212, 213, 217, 218, 222, 233, 234, 235, 236, 237, 238, 241, 242, 243, 244, 245, 246, 252, 253, 255, 257, 258, 259, 260, 261, 274, 275, 279
flying squad 33, 34, 46, 47, 49, 56, 60, 177
fog attack 190, 191
freshwater hose reel system 109, 161, 274
fuel 44, 65, 72, 129, 209, 212, 246, 253, 255
fuel spill 84, 85, 157, 233, 235, 259, 265
fuel station fires 253
fully-developed fire stage 174

G

gases 70
gas free engineering 31
growth stage 174

H

H2S 70, 71, 276, 284
halon 1211 78, 79, 128, 172, 237, 248, 249, 250, 251, 252
halon 1301 11, 78, 79, 80, 81, 128, 129, 130, 131, 132, 134, 135, 137, 184, 202, 275, 282
hangar deck fires 252
hangar deck officer (HDO) 206
HCl 70, 201, 228, 284
HCN 70, 201, 228, 284
HDO 208, 219, 260
heat 4, 64, 69, 72, 73, 275
helicopter control officer (hco) 206
helicopter fires 257
heptafluoropropane 75, 80
HFP 67, 75, 80, 81, 125, 128, 129, 130, 131, 132, 133, 134, 135, 136, 137, 170, 178, 179, 181, 184, 185, 199, 201, 202, 275, 285
high temperature alarm systems 102
horizontal loop system 103
hose handling 185
hoseman 54
hose team leaders 213, 245, 257, 258
hydrazine 226, 231, 276
hydrogen chloride 70, 201, 284
hydrogen cyanide 70, 201, 284
hydrogen sulfide 70, 124, 284

I

iet 21, 22, 33, 46, 60, 177, 276, 285
indirect attack 175, 188, 189, 192, 193, 194, 195, 196
inport emergency teams 46
installed carbon dioxide (CO_2) flooding systems 124
integrity watch officer (IWO) 206
internal engine fires 250
investigators 52, 179, 276
IWO 206, 209, 285

J

jettison 207, 220, 221, 245, 246, 258

K

K-90 talisman 94

L

lower explosive limit (LEL) 173, 272

M

magazine sprinkler systems 108, 277
man in charge at the scene 150
manual dry type sprinkler system 146
manual operated switches 103
medical personnel 213, 218
messenger 59, 218, 268
MFFV 213, 219, 220, 222, 233, 237, 238, 242, 244, 245, 253, 256, 260, 277, 285
miscellaneous seawater sprinkler systems 145, 277
missile gas system 162
mobile firefighting vehicle 219

N

NFTI 94, 151, 182, 186, 193, 199, 285
nozzleman 53

O

officer of the deck 22
OJT 10, 13, 285
on-scene leader 51
on-the-job training 9, 10, 210, 221, 265, 285
OOD 22, 23, 32, 59, 103, 197, 286
ordnance cooling 229
ordnance disposal team 45
ordnance handling officer (OHO) 206
OSL 7, 17, 23, 49, 50, 51, 52, 53, 54, 55, 56, 151, 169, 179, 180, 181, 183, 184, 188, 197, 198, 208, 244, 245, 246, 252, 255, 256, 257, 258, 259, 260, 278, 285
overhaul 199, 212, 216
oxygen 66, 68, 71, 74, 226, 228

P

P-25 238, 277
P-100 Portable Pump 11
PAIU 153, 154, 155, 156, 158, 159, 279, 286
PFFPE 239, 240, 242, 245, 250, 254, 255, 278, 286
PKP 67, 68, 75, 80, 82, 83, 84, 85, 110, 170, 172, 221, 230, 231, 237, 238, 248, 249, 250, 251, 252, 256, 271, 279, 286
plugman 55
portable AFFF injection unit 158, 279
portable electric submersible pump 162
portable fire extinguishers 82
portable in-line educator 96
post-fire investigation 202, 279
potassium bicarbonate 75, 80
proximity firefighting protective ensemble 239

Q

Quantab Chloride Titrator 124

R

radiation 6, 72, 94, 275
radioactive materials 232
ramfan 99, 200, 271
rapid response team 150
reflash 178, 198, 199
refractometer 123
repair party leader 50
repair party manual 10, 12, 181
RPL 36, 44, 50, 53, 286

S

SCBA 50, 51, 52, 53, 54, 58, 88, 89, 91, 92, 100, 182, 183, 184, 197, 198, 217, 220, 239, 240, 249, 252, 280, 286
SCBA coordinator 58
scene leader 7, 17, 23, 49, 51, 213, 214, 215, 216, 217, 218, 219, 254, 275, 278, 285
seawater flooding and sprinkler systems 160
self-contained breathing apparatus 88, 280
single-main system 103, 104
smoke 1, 57, 69, 89, 180, 200, 271
smoke controlman 57
solenoid-operated pilot valve 119
SOPV 113, 115, 116, 117, 118, 119, 120, 280, 286

SSN-21 class firemain system 154
SSN-774 class firemain system 154
standard navy fire hose 95
stretcher-bearer 59, 281

T

team leader 52
the effects of fire 68
thermal imager 52, 70, 94, 216, 230, 285, 286
trim system 153, 281

U

upper explosive limit (UEL) 173, 272

V

vari-nozzle 95, 96, 106, 163, 187, 191, 194, 195, 236
ventilation systems 12, 73, 74, 133, 136, 163, 200
vertical offset loop system 103

W

water 75, 76, 98, 113, 161, 193, 282
water mist systems 137
weapons personnel 218, 229, 245, 258, 259
weapons staging area Fires 259

X

XO 17, 18, 21, 22, 24, 151, 286

www.ingramcontent.com/pod-product-compliance
Lightning Source LLC
Chambersburg PA
CBHW071330210326
41597CB00015B/1398
9780578441627